Naturalization

A Physics Theory On Creation

Riley Miller

Naturalization √

By: Riley Parker Miller

Naturalization

FIRST CALL
For the science community

THANK YOU'S

To these individuals and people who never
thought I could publish, but now realize
that I can.

Table of Contents

Introduction: Being can become capable of being what being is capable of that God is that is about how God is about how God is about how God is about how God is about how God is about is about how God is about how good is about good that is about good.

Dedicated to those people, who are in science have gone before me.

Naturalization

Chapter 1
Fiction

This is about how the naturalization process works. This is through naturalization that is about naturalization that is about naturalization that is about naturalization that is about naturalization. This is about naturalization that is about naturalization. This is a science of deformation.

This is about naturalization that is about naturalization that is about naturalization that is about naturalization that is about naturalization that is about naturalization that is about naturalization that is about naturalization that is about naturalization that is about naturalization that is about naturalization that is about naturalization that is about naturalization that is about naturalization that is about naturalization that is about naturalization that is about itself.

Naturalization

This is about naturalization. This is about naturalization and how the naturalization of the world is about the naturalization that the world has. This is about the naturalization that is about naturalization that is about naturalization that is about naturalization that is about naturalization that is about naturalization that is about naturalization that is about mankind. This is about naturalization that is about the real naturalization that is about naturalization in which is about naturalization that is about naturalization that is about naturalization that is about naturalization that is about naturalization that is about naturalization that is about naturalization that is about naturalization that is about natural impulses that are about God and man. These are what naturalization that is about naturalization that is about naturalization that is about naturalization that is about naturalization that is about naturalization that is about naturalization that is about naturalization that is about naturalization that is about naturalization that is about naturalization that is about naturalization that is about naturalization that is about naturalization that is about naturalization that is about naturalization that is about naturalization. This is about naturalization that is about nature and man and environment and

Naturalization

mankind and God and naturalization that is about
good. This is about the good in naturalization that
is about the good in naturalization that is about
naturalization that is about naturalization that is
about naturalization that is about naturalization
that is about naturalization that is about
naturalization that is about naturalization that is
about naturalization that is about naturalization
that is about naturalization that is about
naturalization that is about naturalization that is
about naturalization that is about naturalization
that is about naturalization that is about
naturalization that is about nature and man and
mankind and environment and God and
naturalization that is about what naturalization is
about. This is about God and man that are about
naturalization that is about naturalization that is
about naturalization that is about naturalization
that is about naturalization that is about
naturalization that is about naturalization that is
about naturalization that is about naturalization.
This is about naturalization that is about
naturalization that is about naturalization that is
about naturalization that is about naturalization
that is about naturalization that is about
naturalization that is about naturalization that is
about naturalization that is about naturalization
that is about naturalization that is about
naturalization that is about naturalization that is
about naturalization that is about naturalization
that is about naturalization that is about
naturalization that is about naturalization. This is
about naturalization that is about naturalization
that is about naturalization that is about
naturalization that is about naturalization. This is
about naturalization that is about naturalization

Naturalization

that is about naturalization. This is about naturalization. This is about naturalization. This is about naturalization that is about naturalization. This is about naturalization. Naturalization is naturalization as naturalization. This is about naturalization as naturalization as naturalization as naturalization as naturalization that is about naturalization as naturalization as naturalization. This is about naturalization that is about naturalization. This is about naturalization that is about naturalization that is about naturalization that is about naturalization that is about naturalization.

This is about naturalization that is about naturalization that is about naturalization that is about naturalization that is about naturalization. This is about naturalization that is about naturalization that is about naturalization that is about naturalization that is about naturalization that is about naturalization that is about naturalization that is about naturalization that is about naturalization. This is about naturalization

Naturalization

that is about naturalization that is about naturalization that is about naturalization that is about naturalization that is about naturalization that is about naturalization that is about naturalization that is about naturalization that is about naturalization that is about naturalization. This is about naturalization that is about naturalization. This is about naturalization that is about naturalization that is about naturalization that is about naturalization that is about naturalization that is about naturalization that is about naturalization that is about naturalization that is about naturalization that is about naturalization that is about naturalization that is about naturalization that is about naturalization that is about naturalization that is about naturalization that is about naturalization that is about naturalization. This is about natural processes of mankind.

This is about naturalization that is about naturalization that is about naturalization. This is about naturalization which is about natural tendencies to control man. This is about the naturalization of the world. This is how it operates. It has knowledge. This is what controls what man thinks. This is how man thinks. This is in terms of control and power. This is naturalization and how it works. This is through naturalization. This is a process of thinking and words. This is about how naturalization does this. This is through naturalization processes.

This is about naturalization and what it is about is about naturalization. This is about what is about what is about what is about what is about naturalization. This is about mankind that is about naturalization. This is about what naturalization is about and what it is about is naturalization. This is about naturalization and what it sees. It is about the becoming of man. This is about what the becoming of man is about. It is about mankind. This is how naturalization is about how naturalization is about naturalization that is about how naturalization is about naturalization that is about how naturalization is about how naturalization is about naturalization. This is about how naturalization is about how

Naturalization

naturalization is about how naturalization is about
how naturalization is about how naturalization is
about how naturalization is about how
naturalization is about how naturalization is about
how naturalization is about how naturalization is
about how naturalization is about how
naturalization is about how naturalization is about
how naturalization is about how naturalization is
about how naturalization is about how
naturalization is about how naturalization is about
how naturalization is about how naturalization is
about how naturalization is about how
naturalization is about how naturalization is about
how naturalization is about how naturalization is
about how naturalization is about how
naturalization is about how naturalization is about
how naturalization is about how naturalization is
about how naturalization is about how
naturalization is about how naturalization is about
how naturalization is about how naturalization is
about how naturalization is about naturalization.
This is about naturalization. This is how
naturalization is about how naturalization is about
how naturalization is about how naturalization is
about how naturalization is about how
naturalization is about how naturalization is about
how naturalization is about how naturalization is
about how naturalization is about how
naturalization is about how naturalization is about
how naturalization is about becoming. This is
about deformation. This is how deformation is
about becoming. This is about the world that is
about the world that is about how the world is
about how the world is about how the world is
about how the world is about how the world is
about how the world is about how the world is

Naturalization

about the world. This is about deformation. This
is about how deformation works, and how it is
involved. This is with the making of a order.
This is deformation. This is about how
deformation is about how deformation is about
how deformation is about how deformation is
about how deformation is about how deformation
is about how deformation is about how
deformation is about how deformation is about
how deformation is about how deformation is
about how deformation is about how deformation
is about how deformation is about how
deformation is about how deformation is about
how deformation is. It is about deformation. This
is about how deformation is about deformation
that is about how deformation is about how
deformation is about how deformation is about
how deformation is about how deformation is
about how deformation is about how deformation
is about how deformation is about how
deformation is about how deformation is about
how deformation is about how deformation is
about how deformation is about deformation.
This is about the world. This is about the world.
This is about how the world is about how the
world is about how the world is about the world.
This is how the real world is about the world.
This is about deformation. This is about how
deformation is about how deformation is about
how deformation is about how deformation is
about how deformation is about how deformation
is about how deformation is. This is about how
deformation is. This is about naturalization. This
is about naturalization that is about naturalization
that is about naturalization. This is about it. This
is about this. This is about how naturalization is

Naturalization

about how naturalization is about the world. This is about the world. This is about how the world works. This works for the world. This is about how the world works, and how it works, is about the world. This is about how the world is about the world. This is about how the world is about the world. This is about good. This is about good. This is the primitive definition of deformation. This is that the world is good. This is that the world is good. This is about thinking and cognation. This is about how the world works. This is about how naturalization is about how naturalization is about how naturalization is about how naturalization is about how naturalization is about how naturalization is about how naturalization is about how naturalization is about how naturalization is about how naturalization is about how naturalization is about naturalization is about how naturalization is about how naturalization is about how naturalization is about how naturalization is about how naturalization is about how naturalization is about how naturalization is about how naturalization is about how naturalization is about naturalization is about how naturalization is about naturalization is about how naturalization is about how naturalization is about naturalization. This is about how naturalization is about how naturalization is about how naturalization is about how naturalization is about how naturalization is about how naturalization is about how naturalization is about how naturalization is about how naturalization is about how naturalization is about how naturalization is about how naturalization is about how naturalization is about how naturalization is about

Naturalization

how nature is about man is about environment is about mankind is about God is about naturalization. This is about a theory that is proven. It works its way through mystery and amazement. This is about science and math and how probability works. This is through the world of becoming. This becomes man and how he becomes with the world is amazing. This is about becoming. This is about the wonder of the world. This is about obvious conclusions. This thereby goes naturalization. This is about naturalization and how it works. This is about how naturalization is about naturalization that is about naturalization. This is about amazement and possibility. This is about the wonder of science and how it becomes. This is about how the mystery of life finds us. This is with knowledge.

But, this theory baffles man. They think it is a weird theory of modern knowledge. They think it is the root to all evil. But, in thinking, way and form have long before gone unnoticed. This is about the amazement and fascination of knowledge. This is about the ways of man. This is about the amusement of mankind. This is about the reality of how it can become and how it becomes. This is about man and mankind and environment. This is about any way you want to learn. This is about chance and possibility. This is about amazement and amusement. This is about learning new ways to cope with life and mastering it. This is about it.

This is about all possibilities mastered. This is about how we have a story to tell and we share it. This is about all possibilities mastered.

Naturalization

This is about amazement and amusement. This is about all. This is about all possibilities mastered.

It is about mastery.
It is about focus.
Deformation.
Deformation.

The possibilities are endless. This is about open possibility and chance. The possibilities are endless. This is about possibilities and chance that become with one another. This is about chance and possibility. These are endless.

This is about a theory. This is a theory of knowledge. This is endless. This is about the endless possibilities. The endless frontier is about knowledge and gain. This is about gain and possibilities. This is how deformation does form. This is of knowledge. This is about how knowledge works that is endless. This is about possibility.

This is about endless. This is how possibility works. This is through endless possibility. This is how chance and possibility are. This is about endless chance and possibility. This is about endless chance and possibility. This is about endless. This is about an end. There has to be an end to something and this is it.

This is deformation. This is the art of deformation. This is about chance and possibility. This is the art of becoming. This is how becoming and knowing are. This is how the chance and possibility are. This is in the world

Naturalization

that we have it. It is about chance and possibility that are endless. This is about how chance and possibility are endless. This is about how chance and possibility are endless. This is about how chance and possibility are. This is about an end. This is to all of mankind. This is to the world that we are endless. This is about the world and how it is endless. This is about endless possibility and chance. This is about an open world with endless possibilities. This is about an end that is near. This is about open possibilities and how an end is not near.

This is about possibilities and how they are endless. This is about deformation. This is how deformation is an art-form. This is about deformation and how it is an art-form. This is about open possibilities and how they are endless. This is about how deformation is about endless possibilities and how they are open. This is about deformation. This is about becoming and becoming. This is about an open invitation to become. This is one with God.

This is to all things that you like. This is if you are sane and responsible. This is about becoming. This is about deformation. This is about art and deformation. This is about deformation. This is about deformation. This is about the art-form of deformation. This is about becoming. This is how deformation is becoming. This is about becoming. This is about art and science. This is about believing and achieving. This is about becoming and how we are predestined. This is about math and science. This is about the art-form of deformation. This is

Naturalization

about how we are becoming. We are about becoming because we are becoming. This is about art-form and circumstance. This is about how we learn and how we are capable of it. This is about the art-form of deformation and becoming. This is about how to learn a lesson and how it is learned. This is about possibilities and how they are endless. This is about a word known as deformation. This is about science and math. This is about thought-forming and thought producing. This is about talent and skill. This is about deformation and how it is all of it. This is about how deformation is becoming and becoming is deformation. This is about the customs and practices of our world. This is about how our world copes with it and how it is a success. This is about deformation. This is about how deformation teaches us. This is about through becoming. This is about art-form and practice. This is about the almighty way of living. This is about true nature. This is about how naturalization is about excelling and the wisdom of the world. This is about how excelling and the wisdom becomes what it becomes. This is about excelling and wisdom that excels and has wisdom. This is not philosophy of the world. This is a science of becoming that excels. It excels as much as it is. It is excelling. It is about the becoming of the world and how it excels. It is about how much the world is about how much the world is like how much the world is like how much the world is like how much the world is like how much the world is like how much the world is like how much the world is like how much the world is like how much the world is like how much the world is like how much the world is like how much the world is like

Naturalization

how much the world is like how much the world
is like how much the world is like how much the
world is like how much the world is like how
much the world is like how much the world is like
how much the world is like how much the world
is like how much the world is like how much the
world is like how much the world is like how
much the world is like how much the world is like
how much the world is like how much the world
is like how much the world is like how much the
world is like how much the world is like how
much the world is like how much the world is like
how much the world is like how much the world
is like how much the world is like how much the
world is like how much the world is like how
much the world is like how much the world is like
how much the world is like how much the world
is like how much the world is like how much the
world is like how much the world is like how
much the world is like how much the world is like
how much the world is like how much the world
is like how much the world is like how much the
world is like how much the world is like how
much the world is like how much the world is like
how much the world is like how much the world
is like how much the world is like how much the
world is like how much the world is like how
much the world is like how much the world is like
how much the world is like how much the world
is like how much the world is like how much the
world is like how much the world is like how
much the world is like how much the world is like
how much the world is like how much the world
is like how much the world is like how much the
world is like how much the world is like how
much the world is like how much the world is like

Naturalization

how much the world is like how much the world
is like how much the world is like how much the
world is like how much the world is like how
much the world is like how much the world is like
how much the world is like how much the world
is like how much the world is like how much the
world is like how much the world is like how
much the world is like how much the world is like
how much the world is like how much the world
is like how much the world is like how much the
world is like how much the world is like how
much the world is like how much the world is like
how much the world is like how much the world
is like how much the world is like how much the
world is like how much the world is like how
much the world is like how much the world is like
how much the world is like how much the world
is like how much the world is like how much the
world is like how much the world is like how
much the world is like how much the world is like
how much the world is like how much the world
is like how much the world is like how much the
world is like how much the world is like how
much the world is like how much the world is like
how much the world is like how much the world
is like how much the world is like how much the
world is like how much the world is like how
much the world is like how much the world is like
how much the world is like how much the world
is like how much the world is like how much the
world is like how much the world is like how
much the world is like how much the world is like

Naturalization

how much the world is like how much the world
is like how much the world is like how much the
world is like how much the world is like how
much the world is like how much the world is like
how much the world is like how much the world
is like how much the world is like how much the
world is like how much the world is like how
much the world is like how much the world is like
how much the world is like how much the world
is like how much the world is like how much the
world is like how much the world is like how
much the world is like how much the world is like
how much the world is like how much the world
is like how much the world is like how much the
world is like how much the world is like how
much the world is like how much the world is like
how much the world is like how much the world
is like how much the world is like how much the
world is like how much the world is like how
much the world is like how much the world is like
how much the world is like how much the world
is like how much the world is like how much the
world is like how much the world is like how
much the world is like how much the world is like
how much the world is like how much the world
is like how much the world is like how much the
world is like how much the world is like how
much the world is like how much the world is like
how much the world is like how much the world
is like how much the world is like how much the
world is like how much the world is like how
much the world is like how much the world is like
how much the world is like how much the world
is like how much the world is like how much the
world is like how much the world is like how
much the world is like how much the world is like

Naturalization

how much the world is like how much the world
is like how much the world is like how much the
world is like how much the world is like how
much the world is like how much the world is like
how much the world is like how much the world
is like how much the world is like how much the
world is like how much the world is like how
much the world is like how much the world is like
how much the world is like how much the world
is like how much the world is like how much the
world is like how much the world is like how
much the world is like how much the world is like
how much the world is like how much the world
is like how much the world is like how much the
world is like how much the world is like how
much the world is like how much the world is like
how much the world is like how much the world
is like how much the world is like how much the
world is like how much the world is like how
much the world is like how much the world is like
how much the world is like how much the world
is like how much the world is like how much the
world is like how much the world is like how
much the world is like how much the world is like
how much the world is like how much the world
is like how much the world is like how much the
world is like how much the world is like how
much the world is like how much the world is like
how much the world is like how much the world
is like how much the world is like how much the
world is like how much the world is like how
much the world is like how much the world is like
how much the world is like how much the world
is like how much the world is like how much the
world is like how much the world is like how
much the world is like how much the world is like

Naturalization

how much the world is like how much the world
is like how much the world is like how much the
world is like how much the world is like how
much the world is like how much the world is like
how much the world is like how much the world
is like how much the world is like how much the
world is like how much the world is like how
much the world is like how much the world is like
how much the world is like how much the world
is like how much the world is like how much the
world is like how much the world is like how
much the world is like how much the world is like
how much the world is like how much the world
is like how much the world is like how much the
world is like how much the world is like how
much the world is like how much the world is like
how much the world is like how much the world
is like how much the world is like how much the
world is like how much the world is like how
much the world is like how much the world is like
how much the world is like how much the world
is like how much the world is like how much the
world is like how much the world is like how
much the world is like how much the world is like
how much the world is like how much the world
is like how much the world is like how much the
world is like how much the world is like how
much the world is like how much the world is like
how much the world is like how much the world
is like how much the world is like how much the
world is like how much the world is like how
much the world is like how much the world is like

Naturalization

how much the world is like how much the world
is like how much the world is like how much the
world is like how much the world is like how
much the world is like how much the world is like
how much the world is like how much the world
is like how much the world is like how much the
world is like how much the world is like how
much the world is like how much the world is like
how much the world is like how much the world
is like how much the world is like how much the
world is like how much the world is like how
much the world is like how much the world is like
how much the world is like how much the world
is like how much the world is like how much the
world is like how much the world is like how
much the world is like how much the world is like
how much the world is like how much the world
is like how much the world is like how much the
world is like how much the world is like how
much the world is like how much the world is like
how much the world is like how much the world
is like how much the world is like how much the
world is like how much the world is like how
much the world is like how much the world is like
how much the world is like how much the world
is like how much the world is like how much the
world is like how much the world is like how
much the world is like how much the world is like
how much the world is like how much the world
is like how much the world is like how much the
world is like how much the world is like how
much the world is like how much the world is like
how much the world is like how much the world
is like how much the world is like how much the
world is like how much the world is like how
much the world is like how much the world is like

Naturalization

how much the world is like how much the world
is like how much the world is like how much the
world is like how much the world is like how
much the world is like how much the world is like
how much the world is like how much the world
is like how much the world is like how much the
world is like how much the world is like how
much the world is like how much the world is like
how much the world is like how much the world
is like how much the world is like how much the
world is like how much the world is like how
much the world is like how much the world is like
how much the world is like how much the world
is like how much the world is like how much the
world is like how much the world is like how
much the world is like how much the world is like
how much the world is like how much the world
is like how much the world is like how much the
world is like how much the world is like how
much the world is like how much the world is like
how much the world is like how much the world
is like how much the world is like how much the
world is like how much the world is like how
much the world is like how much the world is like
how much the world is like how much the world
is like how much the world is like how much the
world is like how much the world is like how
much the world is like how much the world is like
how much the world is like how much the world
is like how much the world is like how much the
world is like how much the world is like how
much the world is like how much the world is like
how much the world is like how much the world
is like how much the world is like how much the
world is like how much the world is like how
much the world is like how much the world is like

Naturalization

how much the world is like how much the world
is like how much the world is like how much the
world is like how much the world is like how
much the world is like how much the world is like
how much the world is like how much the world
is like how much the world is like how much the
world is like how much the world is like how
much the world is like how much the world is like
how much the world is like how much the world
is like how much the world is like how much the
world is like how much the world is like how
much the world is like how much the world is like
how much the world is like how much the world
is like how much the world is like how much the
world is like how much the world is like how
much the world is like how much the world is like
how much the world is like how much the world
is like how much the world is like how much the
world is like how much the world is like how
much the world is like how much the world is like
how much the world is like how much the world
is like how much the world is like how much the
world is like how much the world is like how
much the world is like how much the world is like
how much the world is like how much the world
is like how much the world is like how much the
world is like how much the world is like how
much the world is like how much the world is like
how much the world is like how much the world
is like how much the world is like how much the
world is like how much the world is like how
much the world is like how much the world is like

Naturalization

how much the world is like how much the world
is like how much the world is like how much the
world is like how much the world is like how
much the world is like how much the world is like
how much the world is like how much the world
is like how much the world is like how much the
world is like how much the world is like how
much the world is like how much the world is like
how much the world is like how much the world
is like how much the world is like how much the
world is like how much the world is like how
much the world is like how much the world is like
how much the world is like how much the world
is like how much the world is like how much the
world is like how much the world is like how
much the world is like how much the world is like
how much the world is like how much the world
is like how much the world is like how much the
world is like how much the world is like how
much the world is like how much the world is like
how much the world is like how much the world
is like how much the world is like how much the
world is like how much the world is like how
much the world is like how much the world is like
how much the world is like how much the world
is like how much the world is like how much the
world is like how much the world is like how
much the world is like how much the world is like
how much the world is like how much the world
is like how much the world is like how much the
world is like how much the world is like how
much the world is like how much the world is like
how much the world is like how much the world
is like how much the world is like how much the
world is like how much the world is like how
much the world is like how much the world is like

Naturalization

how much the world is like how much the world
is like how much the world is like how much the
world is like how much the world is like how
much the world is like how much the world is like
how much the world is like how much the world
is like how much the world is like how much the
world is like how much the world is like how
much the world is like how much the world is like
how much the world is like how much the world
is like how much the world is like how much the
world is like how much the world is like how
much the world is like how much the world is like
how much the world is like how much the world
is like how much the world is like how much the
world is like how much the world is like how
much the world is like how much the world is like
how much the world is like how much the world
is like how much the world is like how much the
world is like how much the world is like how
much the world is like how much the world is like
how much the world is like how much the world
is like how much the world is like how much the
world is like how much the world is like how
much the world is like how much the world is like
how much the world is like how much the world
is like how much the world is like how much the
world is like how much the world is like how
much the world is like how much the world is like
how much the world is like how much the world
is like how much the world is like how much the
world is like how much the world is like how
much the world is like how much the world is like

Naturalization

how much the world is like how much the world
is like how much the world is like how much the
world is like how much the world is like how
much the world is like how much the world is like
how much the world is like how much the world
is like how much the world is like how much the
world is like how much the world is like how
much the world is like how much the world is like
how much the world is like how much the world
is like how much the world is like how much the
world is like how much the world is like how
much the world is like how much the world is like
how much the world is like how much the world
is like how much the world is like how much the
world is like how much the world is like how
much the world is like how much the world is like
how much the world is like how much the world
is like how much the world is like how much the
world is like how much the world is like how
much the world is like how much the world is like
how much the world is like how much the world
is like how much the world is like how much the
world is like how much the world is like how
much the world is like how much the world is like
how much the world is like how much the world
is like how much the world is like how much the
world is like how much the world is like how
much the world is like how much the world is like
how much the world is like how much the world
is like how much the world is like how much the
world is like how much the world is like how
much the world is like how much the world is like

Naturalization

how much the world is like how much the world
is like how much the world is like how much the
world is like how much the world is like how
much the world is like how much the world is like
how much the world is like how much the world
is like how much the world is like how much the
world is like how much the world is like how
much the world is like how much the world is like
how much the world is like how much the world
is like how much the world is like how much the
world is like how much the world is like how
much the world is like how much the world is like
how much the world is like how much the world
is like how much the world is like how much the
world is like how much the world is like how
much the world is like how much the world is like
how much the world is like how much the world
is like how much the world is like how much the
world is like how much the world is like how
much the world is like how much the world is like
how much the world is like how much the world
is like how much the world is like how much the
world is like how much the world is like how
much the world is like how much the world is like
how much the world is like how much the world
is like how much the world is like how much the
world is like how much the world is like how
much the world is like how much the world is like
how much the world is like how much the world
is like how much the world is like how much the
world is like how much the world is like how
much the world is like how much the world is like
how much the world is like how much the world
is like how much the world is like how much the
world is like how much the world is like how
much the world is like how much the world is like

Naturalization

how much the world is like how much the world
is like how much the world is like how much the
world is like how much the world is like how
much the world is like how much the world is like
how much the world is like how much the world
is like how much the world is like how much the
world is like how much the world is like how
much the world is like how much the world is like
how much the world is like how much the world
is like how much the world is like how much the
world is like how much the world is like how
much the world is like how much the world is like
how much the world is like how much the world
is like how much the world is like how much the
world is like how much the world is like how
much the world is like how much the world is like
how much the world is like how much the world
is like how much the world is like how much the
world is like how much the world is like how
much the world is like how much the world is like
how much the world is like how much the world
is like how much the world is like how much the
world is like how much the world is like how
much the world is like how much the world is like
how much the world is like how much the world
is like how much the world is like how much the
world is like how much the world is like how
much the world is like how much the world is like
how much the world is like how much the world
is like how much the world is like how much the
world is like how much the world is like how
much the world is like how much the world is like
how much the world is like itself. This is about
how much the world is like how the world is like
how the world is like how the world is like. This
is about deformation and how it is about
deformation. This is about how deformation that
is about art. This is about science. This is about
how the art of deformation is about the art of
deformation that is about the art of deformation.

Naturalization

This is about deformation and the art of it all. This is about the usual life. This is about deformation. This is about how deformation is about deformation. This is about how deformation is about deformation is about deformation is about deformation that is about deformation. This is about deformation is about deformation that is about deformation that is about deformation is about how deformation is about

Naturalization

how deformation is about how deformation is
about how deformation is about how deformation
is about how deformation is about how
deformation is about how deformation is about
how deformation is about how deformation is
about how deformation is about how deformation
is about how deformation is about how
deformation is about how deformation is about
how deformation is about how deformation is
about how deformation is about how deformation
is about how deformation is about how
deformation is about how deformation is about
how deformation is about how deformation is
about how deformation is about how deformation
is about how deformation is about how
deformation is about how deformation is about
how deformation is about how deformation is
about how deformation is about how deformation
is about how deformation is about how
deformation is about how deformation is about
how deformation is about how deformation is
about how deformation is about how deformation
is about how deformation is about how
deformation is about how deformation is about
how deformation is about how deformation is
about how deformation is about how deformation
is about how deformation is about how
deformation is about how deformation is about
how deformation is about how deformation is
about how deformation is about how deformation
is about how deformation is about how
deformation is about how deformation is about
how deformation is about how deformation is
about how deformation is about how deformation
is about how deformation is about how
deformation is about how deformation is about

Naturalization

how deformation is about how deformation is
about how deformation is about how deformation
is about how deformation is about how
deformation is about how deformation is about
how deformation is about how deformation is
about how deformation is about how deformation
is about how deformation is about how
deformation is about how deformation is about
how deformation is about how deformation is
about how deformation is about how deformation
is about how deformation is about how
deformation is about how deformation is about
how deformation is about how deformation is
about how deformation is about how deformation
is about how deformation is about how
deformation is about how deformation is about
how deformation is about how deformation is
about how deformation is about how deformation
is about how deformation is about how
deformation is about how deformation is about
how deformation is about how deformation is
about how deformation is about how deformation
is about how deformation is about how
deformation is about how deformation is about
how deformation is about how deformation is
about how deformation is about how deformation
is about how deformation is about how
deformation is about how deformation is about
how deformation is about how deformation is
about how deformation is about how deformation
is about how deformation is about how
deformation is about how deformation is about
how deformation is about how deformation is
about how deformation is about how deformation
is about how deformation is about how
deformation is about how deformation is about

Naturalization

how deformation is about how deformation is
about how deformation is about how deformation
is about how deformation is about how
deformation is about how deformation is about
how deformation is about how deformation is
about how deformation is about how deformation
is about how deformation is about how
deformation is about how deformation is about
how deformation is about how deformation is
about how deformation is about how deformation
is about how deformation is about how
deformation is about how deformation is about
how deformation is about how deformation is
about how deformation is about how deformation
is about how deformation is about how
deformation is about how deformation is about
how deformation is about how deformation is
about how deformation is about how deformation
is about how deformation is about how
deformation is about how deformation is about
how deformation is about how deformation is
about how deformation is about how deformation
is about how deformation is about how
deformation is about how deformation is about
how deformation is about how deformation is
about how deformation is about how deformation
is about how deformation is about how
deformation is about how deformation is about
how deformation is about how deformation is
about how deformation is about how deformation
is about how deformation is about how
deformation is about how deformation is about
how deformation is about how deformation is
about how deformation is about how deformation
is about how deformation is about how
deformation is about how deformation is about

Naturalization

how deformation is about how deformation is
about how deformation is about how deformation
is about how deformation is about how
deformation is about how deformation is about
how deformation is about how deformation is
about how deformation is about how deformation
is about how deformation is about how
deformation is about how deformation is about
how deformation is about how deformation is
about how deformation is about how deformation
is about how deformation is about how
deformation is about how deformation is about
how deformation is about how deformation is
about how deformation is about how deformation
is about how deformation is about how
deformation is about how deformation is about
how deformation is about how deformation is
about how deformation is about how deformation
is about how deformation is about how
deformation is about how deformation is about
how deformation is about how deformation is
about how deformation is about how deformation
is about how deformation is about how
deformation is about how deformation is about
how deformation is about how deformation is
about how deformation is about how deformation
is about how deformation is about how
deformation is about how deformation is about
how deformation is about how deformation is
about how deformation is about how deformation
is about how deformation is about how
deformation is about how deformation is about
how deformation is about how deformation is
about how deformation is about how deformation
is about how deformation is about how
deformation is about how deformation is about

Naturalization

how deformation is about how deformation is
about how deformation is about how deformation
is about how deformation is about how
deformation is about how deformation is about
how deformation is about how deformation is
about how deformation is about how deformation
is about how deformation is about how
deformation is about how deformation is about
how deformation is about how deformation is
about how deformation is about how deformation
is about how deformation is about how
deformation is about how deformation is about
how deformation is about how deformation is
about how deformation is about how deformation
is about how deformation is about how
deformation is about how deformation is about
how deformation is about how deformation is
about how deformation is about how deformation
is about how deformation is about how
deformation is about how deformation is about
how deformation is about how deformation is
about how deformation is about how deformation
is about how deformation is about how
deformation is about how deformation is about
how deformation is about how deformation is
about how deformation is about how deformation
is about how deformation is about how
deformation is about how deformation is about
how deformation is about how deformation is
about how deformation is about how deformation
is about how deformation is about how
deformation is about how deformation is about
how deformation is about how deformation is
about how deformation is about how deformation
is about how deformation is about how
deformation is about how deformation is about

Naturalization

how deformation is about how deformation is
about how deformation is about how deformation
is about how deformation is about how
deformation is about how deformation is about
how deformation is about how deformation is
about how deformation is about how deformation
is about how deformation is about how
deformation is about how deformation is about
how deformation is about how deformation is
about how deformation is about how deformation
is about how deformation is about how
deformation is about how deformation is about
how deformation is about how deformation is
about how deformation is about how deformation
is about how deformation is about how
deformation is about how deformation is about
how deformation is about how deformation is
about how deformation is about how deformation
is about how deformation is about how
deformation is about how deformation is about
how deformation is about how deformation is
about how deformation is about how deformation
is about how deformation is about how
deformation is about how deformation is about
how deformation is about how deformation is
about how deformation is about how deformation
is about how deformation is about how
deformation is about how deformation is about
how deformation is about how deformation is
about how deformation is about how deformation
is about how deformation is about how
deformation is about how deformation is about
how deformation is about how deformation is
about how deformation is about how deformation
is about how deformation is about how
deformation is about how deformation is about

Naturalization

how deformation is about how deformation is
about how deformation is about how deformation
is about how deformation is about how
deformation is about how deformation is about
how deformation is about how deformation is
about how deformation is about how deformation
is about how deformation is about how
deformation is about how deformation is about
how deformation is about how deformation is
about how deformation is about how deformation
is about how deformation is about how
deformation is about how deformation is about
how deformation is about how deformation is
about how deformation is about how deformation
is about how deformation is about how
deformation is about how deformation is about
how deformation is about how deformation is
about how deformation is about how deformation
is about how deformation is about how
deformation is about how deformation is about
how deformation is about how deformation is
about how deformation is about how deformation
is about how deformation is about how
deformation is about how deformation is about
how deformation is about how deformation is
about how deformation is about how deformation
is about how deformation is about how
deformation is about how deformation is about
how deformation is about how deformation is
about how deformation is about how deformation
is about how deformation is about how
deformation is about how deformation is about
how deformation is about how deformation is
about how deformation is about how deformation
is about how deformation is about how
deformation is about how deformation is about

Naturalization

how deformation is about how deformation is
about how deformation is about how deformation
is about how deformation is about how
deformation is about how deformation is about
how deformation is about how deformation is
about how deformation is about how deformation
is about how deformation is about how
deformation is about how deformation is about
how deformation is about how deformation is
about how deformation is about how deformation
is about how deformation is about how
deformation is about how deformation is about
how deformation is about how deformation is
about how deformation is about how deformation
is about how deformation is about how
deformation is about how deformation is about
how deformation is about how deformation is
about how deformation is about how deformation
is about how deformation is about how
deformation is about how deformation is about
how deformation is about how deformation is
about how deformation is about how deformation
is about how deformation is about how
deformation is about how deformation is about
how deformation is about how deformation is
about how deformation is about how deformation
is about how deformation is about how
deformation is about how deformation is about
how deformation is about how deformation is
about how deformation is about how deformation
is about how deformation is about how
deformation is about how deformation is about
how deformation is about how deformation is
about how deformation is about how deformation
is about how deformation is about how
deformation is about how deformation is about

Naturalization

how deformation is about how deformation is
about how deformation is about how deformation
is about how deformation is about how
deformation is about how deformation is about
how deformation is about how deformation is
about how deformation is about how deformation
is about how deformation is about how
deformation is about how deformation is about
how deformation is about how deformation is
about how deformation is about how deformation
is about how deformation is about how
deformation is about how deformation is about
how deformation is about how deformation is
about how deformation is about how deformation
is about how deformation is about how
deformation is about how deformation is about
how deformation is about how deformation is
about how deformation is about how deformation
is about how deformation is about how
deformation is about how deformation is about
how deformation is about how deformation is
about how deformation is about how deformation
is about how deformation is about how
deformation is about how deformation is about
how deformation is about how deformation is
about how deformation is about how deformation
is about how deformation is about how
deformation is about how deformation is about
how deformation is about how deformation is
about how deformation is about how deformation
is about how deformation is about how
deformation is about how deformation is about

Naturalization

how deformation is about how deformation is
about how deformation is about how deformation
is about how deformation is about how
deformation is about how deformation is about
how deformation is about how deformation is
about how deformation is about how deformation
is about how deformation is about how
deformation is about how deformation is about
how deformation is about how deformation is
about how deformation is about how deformation
is about how deformation is about how
deformation is about how deformation is about
how deformation is about how deformation is
about how deformation is about how deformation
is about how deformation is about how
deformation is about how deformation is about
how deformation is about how deformation is
about how deformation is about how deformation
is about how deformation is about how
deformation is about how deformation is about
how deformation is about how deformation is
about how deformation is about how deformation
is about how deformation is about how
deformation is about how deformation is about
how deformation is about how deformation is
about how deformation is about how deformation
is about how deformation is about how
deformation is about how deformation is about
how deformation is about how deformation is
about how deformation is about how deformation
is about how deformation is about how
deformation is about how deformation is about

Naturalization

how deformation is about how deformation is
about how deformation is about how deformation
is about how deformation is about how
deformation is about how deformation is about
how deformation is about how deformation is
about how deformation is about how deformation
is about how deformation is about how
deformation is about how deformation is about
how deformation is about how deformation is
about how deformation is about how deformation
is about how deformation is about how
deformation is about how deformation is about
how deformation is about how deformation is
about how deformation is about how deformation
is about how deformation is about how
deformation is about how deformation is about
how deformation is about how deformation is
about how deformation is about how deformation
is about how deformation is about how
deformation is about how deformation is about
how deformation is about how deformation is
about how deformation is about how deformation
is about how deformation is about how
deformation is about how deformation is about
how deformation is about how deformation is
about how deformation is about how deformation
is about how deformation is about how
deformation is about how deformation is about
how deformation is about how deformation is
about how deformation is about how deformation
is about how deformation is about how
deformation is about how deformation is about
how deformation is about how deformation is
about how deformation is about how deformation
is about how deformation is about how
deformation is about how deformation is about

Naturalization

how deformation is about how deformation is
about how deformation is about how deformation
is about how deformation is about how
deformation is about how deformation is about
how deformation is about how deformation is
about how deformation is about how deformation
is about how deformation is about how
deformation is about how deformation is about
how deformation is about how deformation is
about how deformation is about how deformation
is about how deformation is about how
deformation is about how deformation is about
how deformation is about how deformation is
about how deformation is about how deformation
is about how deformation is about how
deformation is about how deformation is about
how deformation is about how deformation is
about how deformation is about how deformation
is about how deformation is about how
deformation is about how deformation is about
how deformation is about how deformation is
about how deformation is about how deformation
is about how deformation is about how
deformation is about how deformation is about
how deformation is about how deformation is
about how deformation is about how deformation
is about how deformation is about how
deformation is about how deformation is about
how deformation is about how deformation is
about how deformation is about how deformation
is about how deformation is about how
deformation is about how deformation is about
how deformation is about how deformation is
about how deformation is about how deformation
is about how deformation is about how
deformation is about how deformation is about

Naturalization

how deformation is about how deformation is
about how deformation is about how deformation
is about how deformation is about how
deformation is about how deformation is about
how deformation is about how deformation is
about how deformation is about how deformation
is about how deformation is about how
deformation is about how deformation is about
how deformation is about how deformation is
about how deformation is about how deformation
is about how deformation is about how
deformation is about how deformation is about
how deformation is about how deformation is
about how deformation is about how deformation
is about how deformation is about how
deformation is about how deformation is about
how deformation is about how deformation is
about how deformation is about how deformation
is about how deformation is about how
deformation is about how deformation is about
how deformation is about how deformation is
about how deformation is about how deformation
is about how deformation is about how
deformation is about how deformation is about
how deformation is about how deformation is
about how deformation is about how deformation
is about how deformation is about how
deformation is about how deformation is about
how deformation is about how deformation is
about how deformation is about how deformation
is about how deformation is about how
deformation is about how deformation is about
how deformation is about how deformation is
about how deformation is about how deformation
is about how deformation is about how
deformation is about how deformation is about

Naturalization

how deformation is about how deformation is
about how deformation is about how deformation
is about how deformation is about how
deformation is about how deformation is about
how deformation is about how deformation is
about how deformation is about how deformation
is about how deformation is about how
deformation is about how deformation is about
how deformation is about how deformation is
about how deformation is about how deformation
is about how deformation is about how
deformation is about how deformation is about
how deformation is about how deformation is
about how deformation is about how deformation
is about how deformation is about how
deformation is about how deformation is about
how deformation is about how deformation is
about how deformation is about how deformation
is about how deformation is about how
deformation is about how deformation is about
how deformation is about how deformation is
about how deformation is about how deformation
is about how deformation is about how
deformation is about how deformation is about
how deformation is about how deformation is
about how deformation is about how deformation
is about how deformation is about how
deformation is about how deformation is about
how deformation is about how deformation is
about how deformation is about how deformation
is about how deformation is about how
deformation is about how deformation is about
how deformation is about how deformation is
about how deformation is about how deformation
is about how deformation is about how
deformation is about how deformation is about

Naturalization

how deformation is about how deformation is
about how deformation is about how deformation
is about how deformation is about how
deformation is about how deformation is about
how deformation is about how deformation is
about how deformation is about how deformation
is about how deformation is about how
deformation is about how deformation is about
how deformation is about how deformation is
about how deformation is about how deformation
is about how deformation is about how
deformation is about how deformation is about
how deformation is about how deformation is
about how deformation is about how deformation
is about how deformation is about how
deformation is about how deformation is about
how deformation is about how deformation is
about how deformation is about how deformation
is about how deformation is about how
deformation is about how deformation is about
how deformation is about how deformation is
about how deformation is about how deformation
is about how deformation is about how
deformation is about how deformation is about
how deformation is about how deformation is
about how deformation is about how deformation
is about how deformation is about how
deformation is about how deformation is about
how deformation is about how deformation is
about how deformation is about how deformation
is about how deformation is about how
deformation is about how deformation is about

Naturalization

how deformation is about how deformation is
about how deformation is about how deformation
is about how deformation is about how
deformation is about how deformation is about
how deformation is about how deformation is
about how deformation is about how deformation
is about how deformation is about how
deformation is about how deformation is about
how deformation is about how deformation is
about how deformation is about how deformation
is about how deformation is about how
deformation is about how deformation is about
how deformation is about how deformation is
about how deformation is about how deformation
is about how deformation is about how
deformation is about how deformation is about
how deformation is about how deformation is
about how deformation is about how deformation
is about how deformation is about how
deformation is about how deformation is about
how deformation is about how deformation is
about how deformation is about how deformation
is about how deformation is about how
deformation is about how deformation is about
how deformation is about how deformation is
about how deformation is about how deformation
is about how deformation is about how
deformation is about how deformation is about
how deformation is about how deformation is
about how deformation is about how deformation
is about how deformation is about how
deformation is about how deformation is about
how deformation is about how deformation is
about how deformation is about how deformation
is about how deformation is about how
deformation is about how deformation is about

Naturalization

how deformation is about how deformation is
about how deformation is about how deformation
is about how deformation is about how
deformation is about how deformation is about
how deformation is about how deformation is
about how deformation is about how deformation
is about how deformation is about how
deformation is about how deformation is about
how deformation is about how deformation is
about how deformation is about how deformation
is about how deformation is about how
deformation is about how deformation is about
how deformation is about how deformation is
about how deformation is about how deformation
is about how deformation is about how
deformation is about how deformation is about
how deformation is about how deformation is
about how deformation is about how deformation
is about how deformation is about how
deformation is about how deformation is about
how deformation is about how deformation is
about how deformation is about how deformation
is about how deformation is about how
deformation is about how deformation is about
how deformation is about how deformation is
about how deformation is about how deformation
is about how deformation is about how
deformation is about how deformation is about
how deformation is about how deformation is
about how deformation is about how deformation
is about how deformation is about how
deformation is about how deformation is about
how deformation is about how deformation is
about how deformation is about how deformation
is about how deformation is about how
deformation is about how deformation is about

Naturalization

how deformation is about how deformation is
about how deformation is about how deformation
is about how deformation is about how
deformation is about how deformation is about
how deformation is about how deformation is
about how deformation is about how deformation
is about how deformation is about how
deformation is about how deformation is about
how deformation is about how deformation is
about how deformation is about how deformation
is about how deformation is about how
deformation is about how deformation is about
how deformation is about how deformation is
about how deformation is about how deformation
is about how deformation is about how
deformation is about how deformation is about
how deformation is about how deformation is
about how deformation is about how deformation
is about how deformation is about how
deformation is about how deformation is about
how deformation is about how deformation is
about how deformation is about how deformation
is about how deformation is about how
deformation is about how deformation is about
how deformation is about how deformation is
about how deformation is about how deformation
is about how deformation is about how
deformation is about how deformation is about
how deformation is about how deformation is
about how deformation is about how deformation
is about how deformation is about how
deformation is about how deformation is about

Naturalization

how deformation is about how deformation is
about how deformation is about how deformation
is about how deformation is about how
deformation is about how deformation is about
how deformation is about how deformation is
about how deformation is about how deformation
is about how deformation is about how
deformation is about how deformation is about
how deformation is about how deformation is
about how deformation is about how deformation
is about how deformation is about how
deformation is about how deformation is about
how deformation is about how deformation is
about how deformation is about how deformation
is about how deformation is about how
deformation is about how deformation is about
how deformation is about how deformation is
about how deformation is about how deformation
is about how deformation is about how
deformation is about how deformation is about
how deformation is about how deformation is
about how deformation is about how deformation
is about how deformation is about how
deformation is about how deformation is about
how deformation is about how deformation is
about how deformation is about how deformation
is about how deformation is about how
deformation is about how deformation is about
how deformation is about how deformation is
about how deformation is about how deformation
is about how deformation is about how
deformation is about how deformation is about
how deformation is about how deformation is
about how deformation is about how deformation
is about how deformation is about how
deformation is about how deformation is about

Naturalization

how deformation is about how deformation is
about how deformation is about how deformation
is about how deformation is about how
deformation is about how deformation is about
how deformation is about how deformation is
about how deformation is about how deformation
is about how deformation is about how
deformation is about how deformation is about
how deformation is about how deformation is
about how deformation is about how deformation
is about how deformation is about how
deformation is about how deformation is about
how deformation is about how deformation is
about how deformation is about how deformation
is about how deformation is about how
deformation is about how deformation is about
how deformation is about how deformation is
about how deformation is about how deformation
is about how deformation is about how
deformation is about how deformation is about
how deformation is about how deformation is
about how deformation is about how deformation
is about how deformation is about how
deformation is about how deformation is about
how deformation is about how deformation is
about how deformation is about how deformation
is about how deformation is about how
deformation is about how deformation is about
how deformation is about how deformation is
about how deformation is about how deformation
is about how deformation is about how
deformation is about how deformation is about

Naturalization

how deformation is about how deformation is
about how deformation is about how deformation
is about how deformation is about how
deformation is about how deformation is about
how deformation is about how deformation is
about how deformation is about how deformation
is about how deformation is about how
deformation is about how deformation is about
how deformation is about how deformation is
about how deformation is about how deformation
is about how deformation is about how
deformation is about how deformation is about
how deformation is about how deformation is
about how deformation is about how deformation
is about how deformation is about how
deformation is about how deformation is about
how deformation is about how deformation is
about how deformation is about how deformation
is about how deformation is about how
deformation is about how deformation is about
how deformation is about how deformation is
about how deformation is about how deformation
is about how deformation is about how
deformation is about how deformation is about
how deformation is about how deformation is
about how deformation is about how deformation
is about how deformation is about how
deformation is about how deformation is about
how deformation is about how deformation is
about how deformation is about how deformation
is about how deformation is about how
deformation is about how deformation is about
how deformation is about how deformation is
about how deformation is about how deformation
is about how deformation is about how
deformation is about how deformation is about

Naturalization

how deformation is about how deformation is
about how deformation is about how deformation
is about how deformation is about how
deformation is about how deformation is about
how deformation is about how deformation is
about how deformation is about how deformation
is about how deformation is about how
deformation is about how deformation is about
how deformation is about how deformation is
about how deformation is about how deformation
is about how deformation is about how
deformation is about how deformation is about
how deformation is about how deformation is
about how deformation is about how deformation
is about how deformation is about how
deformation is about how deformation is about
how deformation is about how deformation is
about how deformation is about how deformation
is about how deformation is about how
deformation is about how deformation is about
how deformation is about how deformation is
about how deformation is about how deformation
is about how deformation is about how
deformation is about how deformation is about
how deformation is about how deformation is
about how deformation is about how deformation
is about how deformation is about how
deformation is about how deformation is about
how deformation is about how deformation is
about how deformation is about how deformation
is about how deformation is about how
deformation is about how deformation is about
how deformation is about how deformation is
about how deformation is about how deformation
is about how deformation is about how
deformation is about how deformation is about

Naturalization

how deformation is about how deformation is
about how deformation is about how deformation
is about how deformation is about how
deformation is about how deformation is about
how deformation is about how deformation is
about how deformation is about how deformation
is about how deformation is about how
deformation is about how deformation is about
how deformation is about how deformation is
about how deformation is about how deformation
is about how deformation is about how
deformation is about how deformation is about
how deformation is about how deformation is
about how deformation is about how deformation
is about how deformation is about how
deformation is about how deformation is about
how deformation is about how deformation is
about how deformation is about how deformation
is about how deformation is about how
deformation is about how deformation is about
how deformation is about how deformation is
about how deformation is about how deformation
is about how deformation is about how
deformation is about how deformation is about
how deformation is about how deformation is
about how deformation is about how deformation
is about how deformation is about how
deformation is about how deformation is about
how deformation is about how deformation is
about how deformation is about how deformation
is about how deformation is about how
deformation is about how deformation is about
how deformation is about how deformation is
about how deformation is about how deformation
is about how deformation is about how
deformation is about how deformation is about

Naturalization

how deformation is about how deformation is
about how deformation is about how deformation
is about how deformation is about how
deformation is about how deformation is about
how deformation is about how deformation is
about how deformation is about how deformation
is about how deformation is about how
deformation is about how deformation is about
how deformation is about how deformation is
about how deformation is about how deformation
is about how deformation is about how
deformation is about how deformation is about
how deformation is about how deformation is
about how deformation is about how deformation
is about how deformation is about how
deformation is about how deformation is about
how deformation is about how deformation is
about how deformation is about how deformation
is about how deformation is about how
deformation is about how deformation is about
how deformation is about how deformation is
about how deformation is about how deformation
is about how deformation is about how
deformation is about how deformation is about
how deformation is about how deformation is
about how deformation is about how deformation
is about how deformation is about how
deformation is about how deformation is about
how deformation is about how deformation is
about how deformation is about how deformation
is about how deformation is about how
deformation is about how deformation is about
how deformation is about how deformation is
about how deformation is about how deformation
is about how deformation is about how
deformation is about how deformation is about

Naturalization

how deformation is about how deformation is
about how deformation is about how deformation
is about how deformation is about how
deformation is about how deformation is about
how deformation is about how deformation is
about how deformation is about how deformation
is about how deformation is about how
deformation is about how deformation is about
how deformation is about how deformation is
about how deformation is about how deformation
is about how deformation is about how
deformation is about how deformation is about
how deformation is about how deformation is
about how deformation is about how deformation
is about how deformation is about how
deformation is about how deformation is about
how deformation is about how deformation is
about how deformation is about how deformation
is about how deformation is about how
deformation is about how deformation is about
how deformation is about how deformation is
about how deformation is about how deformation
is about how deformation is about how
deformation is about how deformation is about
how deformation is about how deformation is
about how deformation is about how deformation
is about how deformation is about how
deformation is about how deformation is about
how deformation is about how deformation is
about how deformation is about how deformation
is about how deformation is about how
deformation is about how deformation is about
how deformation is about how deformation is
about how deformation is about how deformation
is about how deformation is about how
deformation is about how deformation is about

Naturalization

how deformation is about how deformation is
about how deformation is about how deformation
is about how deformation is about how
deformation is about how deformation is about
how deformation is about how deformation is
about how deformation is about how deformation
is about how deformation is about how
deformation is about how deformation is about
how deformation is about how deformation is
about how deformation is about how deformation
is about how deformation is about how
deformation is about how deformation is about
how deformation is about how deformation is
about how deformation is about how deformation
is about how deformation is about how
deformation is about how deformation is about
how deformation is about how deformation is
about how deformation is about how deformation
is about how deformation is about how
deformation is about how deformation is about
how deformation is about how deformation is
about how deformation is about how deformation
is about how deformation is about how
deformation is about how deformation is about
how deformation is about how deformation is
about how deformation is about how deformation
is about how deformation is about how
deformation is about how deformation is about
how deformation is about how deformation is
about how deformation is about how deformation
is about how deformation is about how
deformation is about how deformation is about

Naturalization

how deformation is about how deformation is
about how deformation is about how deformation
is about how deformation is about how
deformation is about how deformation is about
how deformation is about how deformation is
about how deformation is about how deformation
is about how deformation is about how
deformation is about how deformation is about
how deformation is about how deformation is
about how deformation is about how deformation
is about how deformation is about how
deformation is about how deformation is about
how deformation is about how deformation is
about how deformation is about how deformation
is about how deformation is about how
deformation is about how deformation is about
how deformation is about how deformation is
about how deformation is about how deformation
is about how deformation is about how
deformation is about how deformation is about
how deformation is about how deformation is
about how deformation is about how deformation
is about how deformation is about how
deformation is about how deformation is about
how deformation is about how deformation is
about how deformation is about how deformation
is about how deformation is about how
deformation is about how deformation is about
how deformation is about how deformation is
about how deformation is about how deformation
is about how deformation is about how
deformation is about how deformation is about
how deformation is about how deformation is
about how deformation is about how deformation
is about how deformation is about how
deformation is about how deformation is about

Naturalization

how deformation is about how deformation is
about how deformation is about how deformation
is about how deformation is about how
deformation is about how deformation is about
how deformation is about how deformation is
about how deformation is about how deformation
is about how deformation is about how
deformation is about how deformation is about
how deformation is about how deformation is
about how deformation is about how deformation
is about how deformation is about how
deformation is about how deformation is about
how deformation is about how deformation is
about how deformation is about how deformation
is about how deformation is about how
deformation is about how deformation is about
how deformation is about how deformation is
about how deformation is about how deformation
is about how deformation is about how
deformation is about how deformation is about
how deformation is about how deformation is
about how deformation is about how deformation
is about how deformation is about how
deformation is about how deformation is about
how deformation is about how deformation is
about how deformation is about how deformation
is about how deformation is about how
deformation is about how deformation is about
how deformation is about how deformation is
about how deformation is about how deformation
is about how deformation is about how
deformation is about how deformation is about
how deformation is about how deformation is
about how deformation is about how deformation
is about how deformation is about how
deformation is about how deformation is about

Naturalization

how deformation is about how deformation is
about how deformation is about how deformation
is about how deformation is about how
deformation is about how deformation is about
how deformation is about how deformation is
about how deformation is about how deformation
is about how deformation is about how
deformation is about how deformation is about
how deformation is about how deformation is
about how deformation is about how deformation
is about how deformation is about how
deformation is about how deformation is about
how deformation is about how deformation is
about how deformation is about how deformation
is about how deformation is about how
deformation is about how deformation is about
how deformation is about how deformation is
about how deformation is about how deformation
is about how deformation is about how
deformation is about how deformation is about
how deformation is about how deformation is
about how deformation is about how deformation
is about how deformation is about how
deformation is about how deformation is about
how deformation is about how deformation is
about how deformation is about how deformation
is about how deformation is about how
deformation is about how deformation is about
how deformation is about how deformation is
about how deformation is about how deformation
is about how deformation is about how
deformation is about how deformation is about
how deformation is about how deformation is
about how deformation is about how deformation
is about how deformation is about how
deformation is about how deformation is about

Naturalization

how deformation is about how deformation is
about how deformation is about how deformation
is about how deformation is about how
deformation is about how deformation is about
how deformation is about how deformation is
about how deformation is about how deformation
is about how deformation is about how
deformation is about how deformation is about
how deformation is about how deformation is
about how deformation is about how deformation
is about how deformation is about how
deformation is about how deformation is about
how deformation is about how deformation is
about how deformation is about how deformation
is about how deformation is about how
deformation is about how deformation is about
how deformation is about how deformation is
about how deformation is about how deformation
is about how deformation is about how
deformation is about how deformation is about
how deformation is about how deformation is
about how deformation is about how deformation
is about how deformation is about how
deformation is about how deformation is about
how deformation is about how deformation is
about how deformation is about how deformation
is about how deformation is about how
deformation is about how deformation is about
how deformation is about how deformation is
about how deformation is about how deformation
is about how deformation is about how
deformation is about how deformation is about

Naturalization

how deformation is about how deformation is
about how deformation is about how deformation
is about how deformation is about how
deformation is about how deformation is about
how deformation is about how deformation is
about how deformation is about how deformation
is about how deformation is about how
deformation is about how deformation is about
how deformation is about how deformation is
about how deformation is about how deformation
is about how deformation is about how
deformation is about how deformation is about
how deformation is about how deformation is
about how deformation is about how deformation
is about how deformation is about how
deformation is about how deformation is about
how deformation is about how deformation is
about how deformation is about how deformation
is about how deformation is about how
deformation is about how deformation is about
how deformation is about how deformation is
about how deformation is about how deformation
is about how deformation is about how
deformation is about how deformation is about
how deformation is about how deformation is
about how deformation is about how deformation
is about how deformation is about how
deformation is about how deformation is about
how deformation is about how deformation is
about how deformation is about how deformation
is about how deformation is about how
deformation is about how deformation is about
how deformation is about how deformation is
about how deformation is about how deformation
is about how deformation is about how
deformation is about how deformation is about

Naturalization

how deformation is about how deformation is
about how deformation is about how deformation
is about how deformation is about how
deformation is about how deformation is about
how deformation is about how deformation is
about how deformation is about how deformation
is about how deformation is about how
deformation is about how deformation is about
how deformation is about how deformation is
about how deformation is about how deformation
is about how deformation is about how
deformation is about how deformation is about
how deformation is about how deformation is
about how deformation is about how deformation
is about how deformation is about how
deformation is about how deformation is about
how deformation is about how deformation is
about how deformation is about how deformation
is about how deformation is about how
deformation is about how deformation is about
how deformation is about how deformation is
about how deformation is about how deformation
is about how deformation is about how
deformation is about how deformation is about
how deformation is about how deformation is
about how deformation is about how deformation
is about how deformation is about how
deformation is about how deformation is about
how deformation is about how deformation is
about how deformation is about how deformation
is about how deformation is about how
deformation is about how deformation is about

Naturalization

how deformation is about how deformation is. It is about deformation.

This is about deformation and how it teaches us to become. This is with the world. This is what the world likes that is about the world. This is about what the world is about and what the world is about is about deformation that is about deformation that is about deformation. This is about deformation and what it is about. This is about deformation that is about deformation which is about deformation. This is about the deformation that is about the world. This is about the world and how the world becomes with heaven. This is about what deformation is about deformation that is about deformation that is about the world. This is about a final conquest of the world and how it is good. This is about good that is about good that is about good. This is about what good is about. This is

Naturalization

about good that is about the good the world has and what it is. This is about how deformation is about how deformation is about how deformation is about deformation that is about the world that is about deformation that is about deformation that is about deformation is about the world that is about the world. This is about deformation that is about deformation that is about deformation that is about deformation that is about deformation that is about deformation that is about all. This is about deformation that is about deformation that is about deformation that is about deformation that is about deformation that is about deformation that is about deformation that is about deformation that is about deformation that is about deformation that is about deformation that is about deformation that is about deformation that is about deformation that is about deformation. This is about the world and how it works. This is about how the world works that is about the world. This is what the world works with. This is about how the world works that is about the world. This is about the world works, that is about the world. This is about the world and the way it works. This is about the world works, and how it works. This is about mankind and how he works. This is through the dilemma of the self. This is how he becomes and becomes. This is about how he becomes. This is about how he becomes and how he becomes is with him. This is about how the world becomes with the world. This is about the world that is about the world. This is about the world that is about the world. This is about the world and the way it looks. This is about the world that is about the world that is about the world. This is about the world that is about the world that is about the

Naturalization

world that is about the world that is about the
world. This is about the world that is about the
world. This is about the world that is about the
world. This is about the world that is about the
world. This is about the world. This is about the
world and how it represents itself. This is through
the art forms of the world. This is about the world
and how it represents itself. This is through the
world. This is about the world that is about the
world and is about the world. This is about what
is about what is about what is about the world.
This is about the world that is about the world.
This is about the world that is about the world.
This is about the world that is about the world that
is about the world. This is about the world is
about that is about the world. This is about the
world. This is about what the world is about.
This is about the world that is about the world.
This is about the world that is about the world.
This is about the world that is about the world.
This is about the world that is about the world.
This is about the world that is about the world.
This is what the world is about and for that is
what the world is about and of. This is about the
world that is about the world. This is about what
the world is about the world that is about the
world. This is about the world that is about the
world that is about the world that is about the
world. This is about the world that is about the
world. This is about the world that is about the
world. This is about the world that is about the
world. This is about the world that is about the
world. This is about the world that is about the
world. This is about the world that is about the
world. This is about the world that is about the
world. This is about the world. This is about the
world that is about the world. This is about the

Naturalization

world. This is what the world is about that is about the world. This is about the world that is about the world. This is about the world that is about the world. This is about the world that is about the world. This is about the world. This is about what the world is about.

Universe.

Universe.

This is about the world that is about the world. This is about the world that is about the world. This is about the world that is about the world. This is what the world is about that is about the world that is about the world. This is about the progress of man. This is about what man became when he became man. This is about the science of becoming. This is about the ways of the universe. Naturalization is a theory of the universe. This is how the universe is about how the universe is about how the universe is about how the universe is about how the universe is about how the universe is about how the universe is about how the universe is about how the universe is about how the universe is about how the universe is about how the universe is about the universe is about how the universe is about how the universe is about how the universe is about how the universe is about how the universe is about how the universe is about how the universe is about how the universe is about how the world is about how the world is about how the world is about how the world is about how the world is about how the universe is about itself. This is about the world. This is about the world that is about the world that is about the world that is about the world that is about the world that is about the world that is about the world. This is about the universe. This is how the universe is

Naturalization

about how the universe is about how the universe
is about how the world is about the world that is
about the world that is about the world that is
about the world that is about the world that is
about the world that is about the world that is
about the world that is about the world that is
about the world that is about the world that is
about the universe that unfolds. This is about the
universe that is about the universe that is about
the universe that is about the world that is about
the world that is about the world that is about the
world that is about the world that is about the
world that is about the universe. This is about the
world that is about the world that is about the
world that is about the world that is about the
world that is about the world that is about the
world that is about the world that is about the
world. This is about the world. This is about the
world that is about the world that is about the
world that is about the world. This is about the
world and how it works. This is about the world
that works that is about the world that works that
is about the world that works. This is about what
the world works with that is what the world
works. This is about how lies and truth are about
the world. This is about the world. This is about
the way the world works. This is about the world.
This is what the world does that is about the
world. This is about what the world works that is
about the world. This is about the world   This is
how the world works for the smartest man ever
Jesus Christ. This is about Jesus Christ that
makes Him the smartest. This is about the
smartest man ever known as Jesus Christ. He is
the King of All Creation that is about Heaven and
Earth. This is about what Jesus Christ stood for.

Naturalization

This is to bring all people to Him and have Him be the beneficiary of all. This is about the world that is about the world. This is about the world that is about the world that is about the world. This is about the world that is about the world that is about the world. Jesus Christ is the smartest man in the world. This is about Jesus Christ and what He is about. He is about love, compassion, care, kindness, natural ability, Christian values, worldly things that really aren't worldly, great things that are meek, great tidings for human beings, virtue, strength, wonderful understanding, goodness, self-control, positive attitude, anything you have been taught in church also. This is about charisma, positive reinforcement, good values, great things that aren't that much cost, good things that have great value to you, the kind of values money can't buy, love, honesty, valuable resources, self-esteem, positive all of it. This is what makes Jesus the smartest man ever. This is because He created religion. This is also what He is. He is God. He is the eternal being not mocked by others. He is the true boss of all. This is how naturalization was created.

This is what God is like in naturalization. This is not a false God in religion. This is not a great teacher. This is what God is that is eternal. This is the God of all. This is what He created. This is the world. This is what God created that is eternal. This is also what God wanted. He wanted to predestine you. So, when people think you are dumb, do not ignore them. Quote a bible verse. This is always for God.

Jesus Christ is the smartest man ever.

Naturalization

Jesus Christ is the smartest man ever, as the smartest man ever, as the smartest man ever. Jesus Christ is the smartest man ever. This is because He was the smartest man ever. This due to the fact that He was the smartest man ever. This is because He was the smartest man ever, that was the smartest man ever, that was the smartest man ever. This is how He was the smartest man ever. He was the smartest man ever, as the smartest man ever, as the smartest man ever. This was because of Him. He was the smartest mane ever, as the smartest man ever, as the smartest man ever. This is because of God. This is what naturalization proves. This is that He was this by natural ability. This was true because of God. This is why God was the smartest man ever. This was because He was Jesus Christ. He was God's Son.

The real God and the real Jesus are God.

This is separate in Christian naturalization and naturalization.

This is why.

This is because God was Jesus also. This is what separates Jesus from God.

This is that Jesus Christ was God's only Son. This is how Christian naturalization works. It works the same as regular naturalization. This is through the process of wisdom. This is why naturalization is about Jesus Christ. This is because Jesus was God and God incarnate. This is how the worldly wisdom separates from the earthly vision of wisdom. This is through the process of naturalization. This is why Jesus became man. This is due to the earthly values of mankind.

Naturalization

The Bible states that this is true. But, if you do not know Christianity, you do not know Jesus in that way. This is what the way that we know Jesus is.

This is through Christian values. This is that He was a good man and God. This is how this proves this. This is through everything. This is how God knows that Jesus is God. This is just by seeing Him. This is through a process known as naturalization also. This is just by seeing. This is obvious that Jesus Christ is God.

But, Christianity has separated that from all of us. This is how Jesus Christ was separated from all of us. This is through the process of becoming.

This is through naturalization. This is how the only way Jesus Christ was Lord is true. This is in a real sense. This is in a scientific manner. This is just that fact is reality in a matter of speaking. This is what reality is. This is just how you believe in it. This is how God intended it to be. This is science as predestination.

This is how God explains His Divine Immaculate Conception Plan. This is through all people. This is why naturalization is His.

This is through a factual, scientific standpoint.

This is through the Christian theory of naturalization.

Chapter 2
Reality
To understand how naturalization works and thinks, one must understand lies and how they work. The way lies work are just by becoming like naturalization does. They become like

becoming works, but differently. They are about
the world that represents them. They are just like
thought and how it works is about how it works in
which is about how it works. This is about things
and how they work. They just appear, and work
differently. They are new knowledge. And, they
have never been created. The lies of the world
become, just by becoming. This is how
naturalization works also. The lies of the world
do not accept naturalization and how they work.
They work through the world that is about the
world that works just how the world works. This
is just about how the world works. This is about
how the world works, which is about how the
reality of the world works. This is about how the
reality of the world works. Lies, work through
reality. This is about reality. This is how reality
works. It is about lies sometimes and how they
are weird. This is about how the reality of the
world and how they work. This is about the
reality of the world and how they become. They
become in fiction. This is about how they are
made up, and do not exist. This is about how lies
exist, and how they exist, is about reality. This is
how we understand the reality of the world, and
how it exists. This is about the reality of the
world. This is about how the world is about
reality. This is about the reality of the world.
This is about the reality of the world that are
about the reality of the world. This is about the
world, that is about the reality of the people. This
is about the reality of the world. This is about the
world, that represents what the world is about.
This is about reality. This is about how the world
works. This is about the world. This is how the
world works. This is about how the world and

Naturalization

how it works. This is about the world, and how it becomes, what is like the world. The world represents itself. This is how lies take shape and form. They are through no format. They do not follow rules. They do not follow existence. They are by itself and nothing else. There is about the change of the world, and they do not embrace it. They do not embrace the world, but only work for evil. The liars in Revelations do go to hell.

Chapter 3
Soul Becoming

This is about naturalization and how it comes to an end. It only ends through Jesus Christ. He is the one who makes all things come to an end. This is just in understanding. Naturalization proves this. This is through a soul search. This would end. All good things must come to an end. This is about how naturalization begins a new chapter. This is about all good things that must come to an end. This is about the Big Bang. This is about science and becoming. This is about how science and becoming don't come to an end. This means that according to naturalization, the planet would end if they had something create it. This is about the theories of the world and how they are false. This is just if you believe this. This is about how the theories are false and how you do not believe in this. This is about soul searching and how this is not it. This is about God and how He is eternal. This is about the will power and how to represent itself. This is through the world. This is by the world. This is by the world. This is perhaps about the

Naturalization

world and perhaps not. This is about the world that is about the world. This is what the world goes by. This is about the world.

Naturalization.

Naturalization.

This is from form and shape. This is about how form and shape work. This is through the matrix. This is the design of the reality of the deformation. This is how deformation works. This is about the world that its design is about and for. This is about how the design and world of the world takes shape. This is through an element known as deformation.

This is how naturalization takes shape. This is how nature and man combine. This is through the process known as deformation. This is about deformation and how it continues and works. This is by a process known as deformation. This is about how deformation works. This is about how deformation works. This is through an element of learning and process. This is about deformation and the art-form. This is about the element of learning and togetherness. This is about the continuance of deformation. It lasts forever. This is about how it lasts and how it lasts is forever. This is because of naturalization.

God is in naturalization. This is how God works and how He works is through deformation. This is about the working of and about and how deformation works. This is through the process of deformation. This is about how it works and how it becomes. This is about how it works and how it works is through deformation. This works through God. This is how God works. It is through a process known as deformation.

Naturalization

This is about how deformation works and how it works is through the continuance of naturalization. This is about how God and man get together. This is about how science and mathematics combine. This is through a theory of God. This is all about naturalization.

But, deformation creates it. It is about the process of continuance and form. This is about shape and form. This is what shape and form are about, and what they are about is substance. This is all in terms of thinking of man. This is about man and how He thinks. This is about mankind and how he thinks. This is about the world and how it relates. This is about the world and how it relates. This is about how the world relates and how it relates is through the world. This is about how the world relates and how it relates is through man. This is about man and environment and about how they relate. This is about man and substance. This is about how to control reality. This is about man and environment and how they relate to each other. This is how man and environment relate. This is about through the world and through the world is how it relates. This is about naturalization. This is how naturalization relates to all. This is through the continuance of mankind in it. This is about how we get along, and how we get along is through progress of man. This is about how the naturalization of the world relates through the naturalization of the world. This is through the continuance of mankind. This is about how the continuance of mankind is about the creation of a theory. This is about how a theory created a world.

Naturalization

This is of naturalization. This is how it created the world. This is through the science of man. This is through becoming. This is what man became and what man became is a natural man. This is through believing in God.

Yet, I do write about the Divine Immaculate Conception Plan, that is about the predestination of mankind. This is about what the world is about that is about what the world is about. This is about the creation of mankind's way of seeing things. This is through a symbol known as naturalization. This is about how naturalization is about what naturalization is about that is about naturalization. This is about naturalization that is about naturalization that is about naturalization that is about naturalization that is about naturalization. This is about a theory of God. This theory explains that God created the Heavens and the earth. This is in this theory. This is about how the world explains this theory. This is through the continuance of man. This is about the continuance of man. This is about man and the continuance of man. This is about the continuance of man. This is through naturalization.

This is through naturalization. This is about how naturalization exists, and how it exists is through naturalization. This is about how God and man relate. This is about how God and man relate. This is about naturalization. This is how naturalization relates with each other. This is through man and environment. This is how naturalization relates to one another. This is about naturalization. This is about naturalization that is about naturalization that is about naturalization. This is about naturalization and

Naturalization

what naturalization sees is about naturalization. This is about the creation of the world. This is about the world that is about the world. This is about the world that is about the world that is about the world. This is about the world and how it is creation. This is about how knowledge is a theory and how it relates to the Bible. This is why it is called science.

Science explains things through the knowledge of man. This is about itself. This is about how science and progress relate. This is through deformation.

This is about the creation of man and how He relates. This is about the creation of God and how He relates. This is about the world and how it relates. This is about how it relates and how the world relates is through the Bible. This is about how the world relates and how the world relates is through the world. This is about how the Bible relates to the world and how it relates to the world. This is how the world relates to the world and how it relates is through the world. This is about how the world relates and how it relates is through the world. This is about creation. This is about mind-solving puzzles of man and God. This is about the world that is about the world. This is about the world and how it relates to man. This is about man and how he relates to the world. This is about the world and how it relates to man. This is about how man relates to the world. This is how the world unfolds and how the world does this is through naturalization. This is in the mind of the student. This is why there is a teacher known as Jesus.

This is about what the world is about and what the world is about is about science. This is

about how the world revolves and how it
becomes. This is through the art-form known as
deformation. This is about the world of
deformation and how the world of deformation is
about the world of deformation. This is about the
world of deformation and how it exists. This is
through the natural ways of man. This is about
the natural way of man. This is about the man
and how it relates. This is through the way of
mankind. This is about how mankind is about
this. This is about how man is about man and is
about man and is about how man is about man.
This is about man and his tale and his story and
his belonging and his gifts and his way of outlook.
This is about how the world is about how the
world is about how the world is about how the
world is about how the world is about how the
world is about how the world is about how the
world is about how the world is about how the
world is about how the world is about how the
world is about how the world is about the world
that is about how the world is about the world is
about how the world is about how the world is
about how the world is about how the world is
about how the world is about the world. This is
about the Bible. This is how the world works that
is about the world that is about how the world is
about how the world is about how the world is
about how the world is about how the world is
about how the world is about how the world is
about how the world is about how the world is
about how the world is about how the world is
about how the world is about how the world is
about how the world is about how the world is
about how the world is about the world.

Naturalization

Chapter 4
Reality bites

This is about how the world becomes. This is through thinking. This is what becomes of man. This is about his mankind and his environment. This is about how stories are. They are about him and him alone.

We live in a fairy tale. This is about loss and gain. It is primarily about loss and gain. This is about belonging and becoming it. This is about thought and the way to overcome it. This is through books.

Books belong to us and we belong to books. This is about believing in philosophy. There is a philosophy behind everything that ever is, was and will be, that is about all that ever is, was, and will be, that is about everything that is,

Naturalization

was, and will be, that is about everything that is
about what is, was, and will be. This is about
problems and solving it. This is about a place in
your mind. This is why you go there. You go
there to learn. This is how the brain works. This
is about through fairy tales and magic. This is
about the deep blue magic of C. S. Lewis. This is
about how he became man.

He was a dinosaur. This is what the book
made me believe. This is in a fictional reality that
wasn't mine. This is how he got there. He did
this by being smart.

When someone gets jealous, they become
smart. This is just what the story made me think.
This is that I was in a fairy tale reality. This is
about being smart and seeing the deep blue magic.
This is a story.
The naturalization of man, is also a fairy tale.
This is about becoming, with God, and with, man.
This is about becoming. This is about how we see
about what is about what is about what is about
what is about what is about what is about what is
about becoming what we have become. This is
about philosophy and science. If we see
something that we believe in, then we do not
believe in it. This is if it is a fairy tale. Some
people just have weird thought. And, this was C.
S. Lewis'.

He was the deep blue magic. This is just
because he wrote about it. This is about
becoming man and how he became man. He did
this to know what the deep blue magic was. All it
was, was story book telling.

This is about the science called
naturalization. This is about how we become with

Naturalization

man. This is about how we do this, and how we do this, is with man. This is how man becomes.

This is with story books. This is about story books and how they relate. They have nothing to do with Christianity. This is why C. S. Lewis was weird and strange.

This is about the theory of naturalization that caused this. This was his book and how it relates. This is just what I thought with a theory known as naturalization. This is because it has always been in me. This is about the reality of known and unknown things. This is about how things become and how things belong. This is about naturalization and how we relate. This is through the reality of the world. This is how we belong, and how we belong, is through theory. This is if you are an individual.

This is only if you are a believer. This is in individual things that are yours. This is how we belong to each other. This is about how belonging to each other is belonging. This is to each other. This is how we belong to each other that is about how we belong to each other. This is about the deep blue magic. This is controversial. This is just to me.

This is how naturalization works. It is through a philosophy known as something like deep blue magic is. This is about becoming and what the deep blue magic is, is not become. This is why it doesn't become. This is because C. S. Lewis was strange.

This is just because my theory doesn't agree with his. This is about what the theory of God is about. This is about the theory of God. He is about theory.

Naturalization

The deep blue magic of C. S. Lewis is about story telling. He tells it like this.

This is what a fairy tale is. It is about the thought of man and how he relates to himself. This is about the material that is about man and God. This is about how man and God exist through becoming. This is about how God and man exist. This is through becoming. This is about how thinking is about how thinking is about how thinking is about how thinking is about how thinking is about how thinking is about how thinking is about how thinking is about how thinking is about how thinking is about how thinking is about how thinking is about how thinking is about how thinking is about how thought is about thought. This is about disorder and the universe. This is about how thinking and the universe, are about. This is about how disorder in the universe is about the universe and how this disorder relates to the universe is about how the disorder relates to the universe, that is about how the disorder in the universe relates to the universe. This is about how disorder relates to the universe. This is through weird and strange things.

This is about how thought and reality form. This is through ways of communication. This is about how ways of communication work, and how they work is through science. This is not what the deep blue magic of C. S. Lewis would believe.

It would believe in one whole world for God. This is about what created the world before God. This is what a supposed statement is. This is also about C. S. Lewis.

Naturalization

This is about how God created the world and how He did it was in seven days.

Science has a hard time believing in it. This is about the way the world was created. People have always speculated about how God or the world was created or created itself. This is behind the deeper meaning of life.

1.   God created man that created this that created this that created this that created this. God created Himself one day. This is because of naturalization. This is only if you believe this. This is a figure of speech. This is because the world was created by God. This is how God created the world.

2.   One day He just decided to create the world. This was from the deep blue magic. This is how He did it. This is by speaking.

3.   This is from a fairy tale. This is about no winning or losing.

4.   God just created the world.

5.   And, from it, He created all.

One day knowledge appeared. This is after He created the world, and after the man ate the apple. This is about how the world was created. This is by the communication of the world. This is through God.

God alone spoke and the world was created. As, for time, He didn't create this. This was what was always around. Time is just a

Naturalization

circle with a triangle in it. This is what some believe. That God is just Himself and Christ is in the middle. This is how fairy tales are.

The naturalization of the world created itself. This is how God just created it. The naturalization of man is a theory after knowledge fell. Adam was like God. This was after he was created. God, of the voice behind creation, spoke and we were created. This is about the world and how it works. It works through the real Bible. This is how God formed Himself. This is through the real world. This is how the real world formed, and how it formed was from a formless void. This was what was before God. No science can explain how God was created because of knowledge.

This is how knowledge formed, that was from creation. This is also God's. So, if we want to know why God created the world, this is just through prayer. This is about how He created the world, and why He created it. This does not explain disorder. If, there was disorder in the universe, then God couldn't have created Himself. This was probably by a divine act of God. This was just to appear. This is about how nature and God created themselves. This was through the trinity also. And, nature of God's is man's human nature. This is what theory really is. It is just a creation of the mind. This is how God created the world. This was through Himself. This was what nature doesn't understand. This is how God created the world. This is why. This is because it was sin nature. This is only after the fall of man. This is because of the fruit of knowledge. If Saint

Naturalization

Thomas Aquinas was right, then God cannot be fathomed or thought of with human words. This is what really is true. This is that God is a phenomenon that doesn't understand Himself. This is what He did. He banned man from the Garden of Eden. This is why we have different knowledge than we did before. This is about how knowledge was about how knowledge was about how knowledge was about how knowledge was about how knowledge was about how knowledge is about how knowledge was about how knowledge was about how knowledge is about how science is about becoming. This is about how knowledge is about how knowledge is about how knowledge is about how knowledge is about how knowledge is about how knowledge is about how knowledge is about how knowledge is about how knowledge is about how knowledge is about how knowledge is about how knowledge is about how knowledge is about how it is really about itself.

This is about how knowledge is about how knowledge is about how knowledge is about how knowledge is about how knowledge is about how knowledge is about knowledge is about how knowledge is about itself.

This is about deformation. This did not create God so no one did. This is about the obvious procedures of naturalization. This is about God and man. All God did was appear one day. This is about how God created the world one day, and how He became. He became with the world. This is about naturalization, and how it is not a false theory. This is just about how God created knowledge through a scientific theory of itself. It is actually a way of creating your own

Naturalization

knowledge. This is through the Bible. This is
about how the Bible was created. This is through
the church. The word known as the Bible was in
the beginning. This was with God and was God.
That was in the beginning. This was what it did.
It created Jesus too, but in our time, it was after
the fall. This is why we believe in Jesus. This is
because of God. He created Jesus to be in the
beginning with us. This was not the deep blue
magic. This was in fairy tales.

This is about deformation. This is a
science and art form. This is about belief and
how it is belief. This is about how belief that is
about God is about man. This is about belief and
how belief shaped our world. This is through
God. This is what didn't create God. This wasn't
faith either. This was just plain old wanting to be
created, so He created Himself. This is about how
God created man. This was in God's image. This
is in the Image of God that God created man.
This was man's image that God created. This was
not Jesus Christ that God created man to be in His
Image. God really does look like man.

But, according to Saint Thomas Aquinas
the world was not even being able to fathom by
man. This is his theory. Mine was that it can be
fathomed by God. This is through us.

This is about how God created the earth.
He created us to think all our thoughts, also. This
is just if we believe this. This means that God is
the one who makes us believe. This is in Him.
This is in the God of the Trinity known as Jesus.
This is not a false God. This is the maker of
Heaven and earth. This is about the earth and
how it was created. This is through God. This is

Naturalization

about how God created the earth. This is through Him. This is how the God of the world created man. This is through all of it. This is through His voice. This is what is a symbol if you want to think of it that way. But, all He did was actually speak. He was this thing known as the world. This is just because He created it. This is how the world was created and how it was created was in the image of God. This is about God and how the world created it. This is about the world and how God created it. This is about how the world created itself. This is just through a theory known as deformation. This is about how the world was created and how it became. This is about how the world was created and how the world was created is about the world. This is about how the world was created. This is about the image of God and how it is created. This is through Jesus also. This is about how God is. He is about God.

     Nietzsche said once that God was dead. This is about what God did and how He did it. He did not know about Nietzsche and He also did not know about war. This is what Nietzsche's motivation was from. He was in the midst of the Nazi's and he was about how he was about how he was about how he was about how he was about how he was about how he was about things that are about him. This is about how Nietzsche was about Nietzsche and this was about Nietzsche. This was about him that was about him. This was about God that was about God. This is that Nietzsche was God and what He wanted was what He got. This was his whole philosophy. This was that God was dead and He was not Him. He was a bad man. This is about his philosophy and what his philosophy was about was about weird things that did not make sense to anyone. This was truth. What he was intending on doing was deceiving. This is why it was false. This is about his life and his works. This is because he was good. This is about the way he was good and how he was good was only in person. This is what an intellectual hypocrite is. It is thinking you are better when you are really worse. This is about how Nietzsche was about man. He was about his thinking that was about him, that was about man. This was a person, as a person, as a person, as a person, as a false prophet. This is about his confusion and his works. This is about how he died. He did this through professing faith in war and not doing what was right.

In this definition he was a hypocrite. This is one who loves falsehood. This is about his teaching and his work. This was bad. This was what no one wanted, but him.

This is why he was a false prophet. This is because he taught wrong. He taught people the wrong idea about God. This is about how the God of Nietzsche was supposedly false. This is about the world and what The Antichrist wants. This is the world. This is what the world wants. It wants itself to be normal. This is about what the world wants, and what the world wants, is

Naturalization

about itself. This is what some can't do about
wrong. This is about the wrong of the life that
was in the best interest of man. This is how man
is about the best interest of man. This is about his
attitude. This is about his attitude and how it
should change. Chuck Colson said that attitude
was everything. This is about the Christian moral
attitude thing. This is about how the world is
ready for Christianity and how the world is ready
for a change. In the words of God, "All things are
possible through Christ who strengthens me."
This is about the Bible and how it is not false. It
is about the world that is about the world that is
about the world. This is about the world. This is
about the world that is about the world that is
about the world that is about the world that is
about the world that is about the world that is
about the world that is about the world that is
about the world that is about the world that is
about the world that is about the world that is
about the world that is about the world that is
about the world that is about the world that is
about the world that is about the world that is
about the world that is about the world that is
about the world that is about the world that is
about the world. This is about the world. This is
about the world that is about the world that is
about the world. This is about the world that is
about the world that is about that is about the
world that is about the world that is about the
world that is about the world that is about the
world. This is about the ways of mankind and the
world. This is about the world and what it is
about. This is about the world and what the world
is about. This is about the world. This is about
the world and what it was about is about the

Naturalization

world. This is about the world and what it is about the world that is about the world that is about the world and the way it represents itself. This is about the world that does things to the world. This is about the world that is about the world that is about the world. This is about the world that is about the world that is about the world that is about the world that is about the world that is about the ways of the world. This is about the world. This is about how the world was created. This is in the image of the world. This is about the world and how it was created. This is about the world and what it represents. This is about the world and how it represents itself. This is through the world and how it represents itself, which is through the world and how it represents itself. This is about the world and how it represents itself. This is about how the world is about itself and how the world represents itself is through the world. This is about the world and what it is through. This is through the world. This is about the world and how it represents itself. This is about how the world represents itself. This is about how the world represents itself. This is about the world and how it represents itself. This is about the world and how it represents itself. This is through the world and how it represents itself.

This is through the world. This is how the world represents itself and what it represents itself means. This is about the world and how it represents itself. This is about the world and how the world represents itself. This is about how the world represents itself and how the world represents itself is through the world. This is about the world and how it works. This is

Naturalization

through the world. This is about the world and
how it represents itself. This is about the world
and how it represents itself. This is about how the
world represents itself. This is through the world.
This is how the world represents itself. This is
through the following of the world. This is about
the following of the world. This is about the
world and how the world is good. This is about
the world and how it represents itself. This is
about itself and how it represents itself. This is
through the world. This is how the world works,
that is about the world. This is about the world
that is about the world that is about the world that
is about the world.   This is about the world and
what it represents. This is about will and how it
represents things. This is about the world. This is
how the world is about how the world is about
how the world is about how the world is about the
world is about how the world is about how the
world is about how the world is about how the
world is about how the world is about how the
world is about how the world is about how the
world is about how the world is about how the
world is about how the world is about how the
world is about how the world is about how the
world is about how the world is about how the
world is about how the world is about how the
world is about how the world is about how the
world is about the world is about how the world is
about how the world is about how the world is
about how the world is about how the world is
about how the world is about how the world is
about how the world is about how the world is
about how the world is about the world. This is
about the world. This is about the world and how
it represents itself. This is about the world that is
about it. This is about it. This is about it. This is

Naturalization

about the world and how it represents itself. This
is through it.

Naturalization is naturalization through it.

It is itself. This is about itself that is about
itself. This is about itself. This is about itself that
is about itself that is about itself that is about
itself. This is through itself. This is about itself.
This is about itself. This is about itself. This is
about itself. This is about itself. This is about
itself. This is about itself. This is about itself.
This is about itself. This is about itself. This is
about itself. This is about itself. This is about
itself. This is about itself. This is about itself.
This is about itself. This is about itself. This is
about itself. This is about itself. This is about
itself. This is about itself. This is about itself.
This is about itself. This is about itself. This is
about itself. This is about itself. This is about
itself. This is about itself. This is about itself.
This is about itself. This is about itself. This is
about itself. This is about itself. This is about
itself. This is about itself. This is about itself.
This is about itself. This is about itself. This is
about itself. This is about itself. This is about
itself. This is about itself. This is about itself.
This is about itself. This is about itself. This is
about itself. This is about itself. This is about
itself. This is about itself. This is about itself.
This is about itself. This is about itself. This is
about itself. This is about itself. This is about
itself. This is about itself. This is about itself.
This is about itself. This is about itself. This is
about itself. This is about itself. This is about

Naturalization

itself. This is about itself. This is about itself.
This is about itself. This is about itself. This is
about itself. This is about itself. This is about
itself. This is about itself. This is about itself.
This is about itself. This is about itself. This is
about itself. This is about itself. This is about
itself. This is about itself. This is about itself.
This is about itself. This is about itself. This is
about itself. This is about the world that is about
the world. This is about what makes it tick. This
is about the world that is about the world. This is
about the world that is about the world. This is
about the world that is about the world. This is
about the world. This is about the world and how
it becomes. This is about the becoming of the
world and how it becomes with the world. This is
about the world and how it becomes itself. This is
about the world and how it becomes itself. This is
about the world and itself. This is about itself.
Deformation is about itself. Deformation is about
itself. This is about deformation. This is about
deformation. This is what deformation is about.
This is what deformation is about. It is about
anything that you want. This is about respect,
authority, and judgment. This is about science
and math and what science and math are about is
about science and math. This is about science and
math. This is what science and math are about.
This is about science and math which are about
science and math. This is about math and science
and about this. This is about itself and what is
about itself is about itself. This is about science
and math. This is about science and math. This is
what science and math are about that are about
science and math. This is about science and math
that are about science and math. This is about

Naturalization

mathematical proportions that are about science and math and what they are about is science and math. This is about mathematics that is about science and math. This is about math and science that is about math and science that is about math and science that is about math and science that is about math and science that is about math and science that are about math and science that are about math and science that are about math and science that are about math and science that are about math and science that are about math and science. This is about math and science that is about math and science that is about math and science that is about math and science. This is about math and science that is about math and science that is about math and science that is about math and science that is about math and science. This is about math and science that is about math and science that is about math and science. This is about math and science that is about math and science that is about math and science that is about math and science that is about math and science that is about math and science that is about math and science that is about math and science that is about math and science that is about math and science that is about math and science that is about math and science that is about the formation of thought and man. This is about man and his thought and what he thinks. This is about math and science. This is how formation of thought according to deformation occurs and happens. This is about a process known as deformation. This is about deformation and what deformation is about is about deformation. This is about deformation that is

Naturalization

about deformation that is about deformation that is about deformation that is about deformation that is about deformation. This is about deformation that is about deformation that is about deformation that is about deformation that is about deformation that is about deformation that is about deformation that is about deformation that is about deformation that is about deformation that is about deformation that is about deformation that is about deformation that is about deformation that is about deformation that is about deformation that is about the world and how it works. This is about the world and how it works. This is about deformation and what deformation is about is about deformation. This is about deformation that is about deformation that is about deformation that is about deformation that is about deformation that is about deformation that is about all. This is about all.

This is about all things that are possible through Christ who strengthens me. This is about deformation and what it means. This is about the natural process of teaching. This is about also the natural process of learning. This is about learning that is about the natural process of mankind. This is about learning and the way learning is about learning. This is about mankind. This is what mankind is about. This is about deformation. This is about learning and how it learns. This is about what learning is about and what learning is about is about deformation. This is about what deformation is about that is about deformation. This is about learning. This is what learning is about that is about deformation. This is about learning the process of deformation. This is about the world and how the world is about is about learning. This is about learning that is about the learning that is about the learning that is about the world that is about learning. This is about the learning that is about the learning that is about the learning that is about the world and its way of learning. This is about the way we learn. This is how we learn. This is about how we learn and how we learn is about how we learn. This is about learning and how we are about it. This is about learning. This is about how we learn that is about learning. This is a process. This is of learning. This is about deformation. This is about how deformation is about how deformation is about how deformation is about how deformation is about how deformation is about deformation is about how deformation is about how deformation is about how deformation is about how deformation is about how deformation is about how deformation that is about how deformation is about how deformation is about how it is about it. It is about it. This is about deformation. This is about how deformation works. It is all about deformation that is about deformation. This is about deformation. This is about deformation that is about order. This is about a reality of good that is about good that is about good that is about good that is about good that is about good. This is about good that is about good that is about good, that is about good. This is about good that is about good that is about good. This is about good that is about good that is about good that is about good. This is about good. This is about good. This is what good is about that is about good. This is about good. This is what good is about. It is about good that is about good.

Naturalization

This is about good. This is about good. This is what good is about that is about good. It is good that is about good that is about good. This is about good that is about good. This is about good that is about good that is about good that is about good. This is about good that is about good that is about good. This is about good that is about good that is about good. This is about good that is about that is about good. This is about good that is about good. This is about good that is about good. This is about good that is about good. This is about good that is about good. This is about good that is about good that is about good. This is about good that is about good. This is about good that is about good that is about good. This is about good that is about good. This is about the world that is about the world. This is about the world and what it is about. This is about the world that is about the world that is about the world. This is about the world and what it is about. This is about the world and how it does it. This is about what the world is about that is about the world. This is about how the world is about how the world is about how the world is about that is about how the world is about how the world is about how the world is about how the world is about that is about how the world is about how the world is about how the world is about how the world is about how the world is about how the world is about the world is about how the world is about how the world is about how the world is about how the world is about how the world is about how the world is about how the world is about how the world is about how the world is about how the world is about how the world is about how the world is about how the world is about how it is about it. This is

Naturalization

about the world. This is about how the world is
about how the world is about how the world is
about how the world is about how the world is
about how the world is about how the world is
about how the world is about how the world is
about how the world is about how the world is
about how the world is about how the world is
about the world. This is how the world is about
how the world is about how the world is about
how the world is about how the world is about
how the world is about how the world is about
how the world is about all. This is about the
world that is about the world that is about the
world that is about the world that is about the
world that is about the world that is about the
world that is about the world that is about the
world that is about the world. This is about the
world that is about the world. This is about the
world. This is about what the world is about that
is about the world. This is what the world wants
that is about the world. This is about the world
that is about the world that is about the world that
is about the world. This is about the world. This
is about the world that is about the world that is
about the world. This is about what the world is
about that is about the world that is about the
world that is about the world. This is about the
world. This is what the world is about that is
about the world that is about the world. This is
about the world. This is what the world is about
that is about what the world is about. This is
about the world. This is about the world that is
about the world. This is about the world that is
about the world. This is about the world. This is
about what the world is about. This is about the
world and what it is about. It is about itself. This

Naturalization

is about itself. This is about the world that is
about the world that is about the world. This is
about the world and about itself. This is about
how the world is about itself. This is about itself.
This is about the world. This is about itself. This
is about itself. This is about itself. This is what
itself is about. It is about itself. This is about
itself. This is about itself. This is what itself is.
This is what itself is. This is about naturalization
and how it is about naturalization   This is about
sin and how it works. This is about the world that
is about the world. This is about the world that is
about the world that is about the world. This is
about the world that is about the world that is
about the world that is about the world. This is
about how the world has sin and what the world is
about sin that is about sin. This is about sin and
how the world is about sin and what the world has
is sin. This is in the world. This is about how the
world is about sin and what sin is about is about
the world. This is a place of inbreeding and hate.
This is about how sin is about how sin is about
how sin is about how sin is about sin that is about
sin. This is about sin that is about sin and what
sin is about is about sin. This is about sin. This is
what sin is about and what sin is about is about
sin. This is about sin and what sin is about is
about sin. This is about what sin is about. This is
about sin that is about sin that is about is about
sin. This is about the reality that is about sin. Sin
is about the reality known as sin and what it is
about is about sin. This is about sin that is about
sin that is about sin. This is about sin and what
sin is about is about sin. This is about sin that is
about sin. This is about sin that is about sin. This
is what sin is about that is about sin. This is about

Naturalization

the way in which sin is about sin. This is about
sin that is about sin. This is about sin that is about
sin. This is about sin that is about sin. This is
about sin that is about sin. This is about sin
which is about the way we sin. This is how sin
works. This is about sin and about how sin works
that is about sin. This is about sin and how sin
works. This is about sin and what sin is about is
about sin. This is about what sin is about that is
about sin. This is about sin which is about sin
that is about sin. This is about sin which is about
sin that is about sin that is about sin. This is about
sin that is about sin. This is about sin that is about
sin. This is what sin is about that is about sin.
This is about sin that is about sin that is about sin.
This is about sin that is about sin. This is about
sin that is about sin that is about sin. This is about
sin that is about sin. This is about sin that is about
sin. This is about sin that is about sin. This is
about sin that is about sin. This is about sin that is
about sin. This is about sin that is about sin. This
is about nature that is about sin. This is about sin
that is about sinning. This is about what sinning
is about that is about sin. This is about how sin
controls people. This is about sin and what sin is
about is about sin. This is what sin is about that is
about sinning. This is about one's nature and how
they sin. This is through the nature of man. This
is about the nature of man. This is how the nature
of mankind is and what it is about is about what it
is about. This is about mankind. This is about
sinning that is about mankind that is about
mankind. This is about mankind. This is about
mankind that is about sinning. This is about his
nature. It really sins. This is about nature that is
about man that is about mankind. This is about

Naturalization

mankind that is about mankind that is about mankind that is about mankind. This is about the world that is about mankind that is about mankind. This is about sin. This is what sin is about. It is about the thing that is about the thing that is about the thing that is about the thing that is about the thing that is about the thing that is about the thing that is about the thing that is about the thing that is about the thing that is about the thing that is about the thing that is about the thing that is about the thing that is about the thing that is sin. This is about the bad of the world. This is what sin is about that is about sin. This is about sin that is about sin. This is what sin is about that is about sin. This is about sin that is about sin that is about sin. This is about sin that is about sin. This is about sin that is about sin. This is about sin that is about sin. This is what sin is about. This is about sin that is about sin. This is about sin that is about sin. This is about sin. This is about sin that is about sin. This is about sin that is about sin. This is about sin that is about sin. This is about sin that is about sin. This is about bad things that are about sin and that are about sin. This is about sin that is about sin. This is about what sin is about that is about sin. This is what sin is about. This is about sin. This is about sin that is about sin. This is about sin that is about sin. This is about sin that is about sin. This is what sin is about that is about sin. This is about sin that is about sin. This is about sin that is about sin. This is about sin that is about sin. This is about sin that is about sin. This is about sin that is about sin. This is about sin that is about sin. This is about sin that is about sin. This is about sin that is about sin. This is what sin is about and

Naturalization

what sin is about is about sin. This is about sin
that is about sin. This is what sin is about that is
about what sin is about that is about what sin is
about that is about what sin is about that is about
what sin is about that is about what sin is about
that is about what sin is about that is about what
sin is about that is about what sin is about that is
about what sin is about that is about what sin is
about that is about what sin is about that is about
what sin is about that is about what sin is about
that is about what sin is about that is about what
sin is about that is about what sin is about that is
about what sin is about that is about what sin is
about that is about what this thing is. It is about
sin. This is about what is about what is about
what is about what is about what is about what is
about what is about what is about what is about
what is about what is about what is about what is
about what is about what is about what is about
sin. This is about sin. This is what sin is about
and what sin is about is about extraordinary
things. This is pure bad. This is how sin works
and how it works is through sin. This is about sin
and what is about it is about itself. Sin is also
only about itself. This is about what sin is about
that is about itself. This is about itself that sin is
about. This is about sin and what it is about is
about sin itself. This is about sin and what sin is
about is really truly what is about sin. This is
about sin. This is what sin is about and what sin
is about is about sin. This is about what sin and
what it is about is about sin. This is about sin.
This is what sin is about. This is about sin and
what it is about is about sin. This is about sin.
This is what sin is about. This is about sin. This
is what sin truly is that is about sin. This is about

Naturalization

sin. This is what sin is about and what sin is about is what sin is about. This is about sin. This is what sin is about that is about is about sin. This is about sin and what it does is unspeakable. This is about what sin is about and how it is about sin is what it is about. This is about sin. This is what sin is about. It is about sin and what sin is about is about sin. This is what sin is about and what sin is about is about sin. This is about sin that is about sin. This is about sin and what sin is about is about sin. This is about sin is about that is about sin. This is about sin and what sin is about is about sin. This is about sin. This is what sin is about and what sin is about is about sin. This is about sin and the ways of sin. This is about what sin is about that is about sin. This is about sin. This is about sin. This is about sin that is about sin. This is about sin that is about sin that is about sin that is about sin. This is about sin. This is about sin and what it is about. This is about sin. This is what sin is about that is about sin. This is about Jesus and how He saves. This is how sin works and what it is about is about sin. This is about sin that is about sin. This is about sin that is about sin. This is about sin that is about sin that is about sin that is about that is about sin. This is about sin that is about sin and what sin is about is about sin. This is about sin and what sin is about is about sin. This is about sin that is about sin. This is what the wages of sin is: death. This is about sin that is about sin that is about sin that is about sin. This is about sin. This is about sin that is about sin that is about sin. This is about sin that is about sin. This is about sin that is about sin that is about sin. This is about sin. This is about what sin is about and what sin is about is about

Naturalization

sin. This is about sin that is about sin. This is
about sin that is about sin that is about sin. This is
about sin that is about sin that is about sin. This is
about sin that is about sin. This is about sin that is
about sin that is about sin. This is about sin that is
about sin that is about sin that is about sin that is
about sin that is about sin that is about sin that is
about sin that is about sin. This is about sin that is
about sin that is about sin that is about sin that is
about sin. This is about sin that is about sin that is
about sin that is about sin that is about sin that is
about sin that is about sin that is about sin that is
about sin that is about sin that is about sin that is
about sin. This is about sin that is about sin that is
about sin that is about sin. This is about sin that is
about sin that is about sin that is about. This is
about sin that is about sin that is about sin that sin
is about sin that is about sin that is about sin. This
is about sin. This is about sin that is about sin that
is about sin. This is about sin that is about sin that
is about sin that is about all. This is about how
sin comes from the mind. This is about how sin
does not come from the mind. It does and it
doesn't. This means that it takes shape there and
manifests itself to the world. This is how clear
thought is. It is without sin. This is why sin is so
bad. This is why the meaning of life is so bad.
This is because sin came from the person that was
life itself. This person was you. This was not
anybody's fault. This is why sin is bad. It is
because the sin of the world is what sin is about.
This is about the sin of the world. This is about
the sin of the world being the sin of the world.
This is about the sin of the world being the sin of
the world. This is about being saved, and how
you can be saved. This is about being saved and

Naturalization

being saved is being about being saved. This is about the thing that has always been. This is ever since Adam ate the apple. There has always been sin. This is in the world. This is about the world that is about the world that is about the world. This is about the world that is about the world that is about the world that is about the world. This is about the world that is about sin and the world. This is about the world that is about the world. This is about the world that is about the world that is about the world. This is about the world that is about the world. This is about what is about the world. This is about the world. This is what the world that is about the world. This is about what the world is about that is about the world. This is about what the world is about that is about the world. This is about the world. This is about the world that is about the world that is about the world. This is about the world that is about the world that is about the world. This is about the world that is about the world that is about the world that is about the world. This is about the world that is about the world is about the world that is about the world. This is about the world that is about the world. This is about the world that is about the world that is about the world that is about the world that is about the world that is about the world. This is about the world that is about the world that is about the world that is about the world that is about the world that is about the world that is about the world that is about the world that is about the world that is about the world that is about the world, that is about the world that is about the world that is about the world that is about the world that is about the world that is about the world that is about the world that is about the world that is

about the world that is about the world. This is
about the world that is about the world that is
about the world that is about the world that is
about the world that is about the world that is
about the world. This is about the world. This is
about the world that is about the world. This is
about the world that is about the world that is
about the world that is about the world that is
about the world that is about the world that is
about the world that is about the world that is
about the world that is about the world that is
about the world that is about the world that is
about the world that is about the world that is
about the world that is about the world that is
about the world that is about the world that is
about the world that is about the world that is
about the world that is about the world that is
about the world that is about the world that is
about the world that is about the world that is
about the world that is about the world that is
about the world that is about the world that is
about the world that is about the world that is
about the world that is about the world that is
about the world that is about the world that is
about the world that is about the world that is
about the world that is about the world that is
about the world. This is about the reality of the
world that is about the reality of the world that is
about the reality of the world that is about the
reality of the world that is about the reality of the
world that is about the reality of the world. This
is about the reality of the world. This is about the
world that is about the world that is about the
world that is about the world that is about the
world that is about the world that is about the
world that is about the world that is about the

Naturalization

world that is about the world that is about the
world that is about that is about the world that is
about the world that is about the world that is
about the world that is about the world that is
about the world that is about the world that is
about the world that is about the world. This is
about the world that is about the world that is
about the world that is about the world that is
about the world that is about the world that is
about the world that is about the world that is
about the world that is about the world that is
about the world that is about the world. This is
about the world that is about the world that is
about the world that is about the world. This is all
good. This is why I say this. This is because of
all the world's mysteries that are ours. They are
ours because they are good. Yet, people are not
always good.

This is because of perception. Perception
is always good. This is why all the mysteries of
the world are ours. This is because sin is one of
them. This is the reason for this. This is due to
the fact that sin is bad because people are bad.
This is why this is bad. This is because it is sin.
This is why it is bad. This is because it is sin.
The world is good. This is real good. This is
because God created the world to be good.

Yet, mankind and the serpent fell. This is
what happened when he fell. He became bad.
Yet, all of man is not bad. This is because of
Jesus. He is the one who made us in His image in
addition to God. This is why He is God also.
This is because the sin of the world is bad. This
means that the sin of the world is about the sin of
the world. This is about why the sin of the world
is bad. This is because it is about the way it is

Naturalization

about itself. This is because of the why. This is because of the what.

Sin is sin as sin. This is because man fell. This is because it is a bad thing. This is the art of sin. This is how sin is not a bad thing. This is because it is the worst. We do not even notice it. This is because of sin. That is what sin is about, that is about what sin is about. It is about bad things. Yet, Jesus came to save. He knew about sin and what it was all about. This is about the opposite of Him. That is why He came to save. It is because of sin we fell. This is about the becoming of man. This is how man becomes. This is by not falling anymore.

This is about a clean heart. This is why a clean heart is for God. It is why it is. Sin is about sin which is about sin. It is why it is around.

The reality of the world of God that was before time was that sin did not create the world. It wasn't around.

The world was created by God. Why He created the world was through Himself. It was to please man. God created the world for man. God created man to please Him. Why He did this was because of man. God created the earth in seven days to please man. God created the earth to please man. The world was created by God, which was created by God, which was created by God, which was created by God, which was created by God, which was created by God, which was created by God. God created the world, which created the world, which created the world, which created the world, which created the world, which created the world, which created the world. He created the world to create the universe. God created the world, to

Naturalization

create the world, to create the world, to create the world. God created the world just to create the world. He created the world for man. God created the world to create it. Sin did not create the world. Sin was just there. It was a tree in the Garden of Eden, which Adam saw. This was the only rule of God.

This was to not sin. This was to not give in. This is what the serpent was tricky enough to make Adam do. This is through temptation. This is why God did not create the world. That is why it looks like this. He created man in His Image. This was the world besides the tree of knowledge. This was God's own tree. This was why it was God's. This is because God made it that way. God created the world and there is no debate about all of everything in which God did. This is why. God created the world. Judgment comes from Jesus alone. This is what created the world. This is when Jesus saw it was good. This is after He came. This is what became of the world. This was an empty, formless void. It doesn't matter if God didn't create it. It matters that He did create all of the world. Evolution would have only created thought. This is why it matters. This is because of corrupt, nonsensical people. They are the ones that do not believe this. And, they become weird because of this. Religion is God's, and God's is religion. This is why we fell.

This is because of Adam alone. If we did not like Adam, that means we are very sane in the ways of all. This is what Adam did, he fell because of the serpent. This is what is cursed the most among all animals. This is why God did it.

He became angry when man fell. And, He ended the reality of the world because of it. God

created the world to end it. So, it is God's Divine
Immaculate Conception. This is to save man.
This is why He created it. This is to educate and
inform. This is about the salvation of man. God
is what is about the plan that is about the plan. It
is God's plan and not ours. God created man to
serve him. He is why there is a master plan. This
is because of Judas. If it weren't for Judas, Jesus
would have had a good life. This is why God fell.
This is because of betrayal. This is why Judas
was crazy. He probably would have believed this
if could have. This is that he was a betrayer. This
is why Judas was a betrayer. It is all about the
Bible. And, he got jealous of this. This is why he
is not in the Bible.

This is about deformation. This leads all
people astray. This is just if you want it to
become in this formation of the way. Jesus is the
way. He is the one true way. This is of believing.
This is why deformation leads people astray. It
all does if you want this to be this way. This is
about believing. This is what people have done.
This is how people believe in people that are
about people. This is about the world that is
about the world that is about the world that is
about the world that is about the world that is
about the universe. This is about the world that is
about God. God is about God. God is about the
way of the world that sis about God. This is about
God that is about the world. This is about how
God created the world. This is how God did it.
He created the Heavens. This is about how the
Bible is of and for God. This is about how the
Bible is about how God is for and of us. This is
about the way the world unfolds. This is about
how the way of the world is about the way of the

Naturalization

world. This is about the way of the world. This is about how the world unfolds and how it unfolds is through the world. This is about the world and how it unfolds. This is through religion. This is about how religion rules the world and how the world rules religion. This is about how the world rules the world and how the world rules the world is through the way of the world. This is about the way of the world. This is about the way of the world and how it is about God. This is about God and how He is about Himself. This is about Himself that is about Himself. This is about Him that is about Himself. This is about Himself that is about Himself, that is about Himself. He is about Himself. This is what is about God that is about Himself. This is about a teaching power of Jesus. This is to teach us everything we believe, and believe in it.

The Antichrist and How Naturalization Becomes with His Knowledge

This is about the Antichrist. He is about the watchful eyes of God. He is Satan, and He is powerful. This is about the way of the world and how it watches us. This is because of The Antichrist.

This is about a power of The Antichrist. He is the watchful deceiver of the world. This is about His attitude and personality. He is an angel that is about being about the ways of the world. He is about being better than people, and about being superior to everyone. This is about lies, and how He is them. This is about an angel and how he thinks. He is about lies and deceit because this is what keeps him alive. He is about shoving and pushing children in public. He is about hating and lusting over life. He is about life, and life abundantly.

He is about the life abundantly that fools people. He is about popularity and how He started it. This is in His terms only. He likes to torture and harass people of low quality. This is after He lies to make them do it. This is about his lies and deceit. This is about the wisdom of The Antichrist and how He works. He works like a child that knows how to push an adult. But, the adult is really pushing the child. He is about stardom and personality. This is about how much he likes this thing. He is about the world and how he treats it, gently. It is a kind of anger he has, that is unstoppable. He is about pushing blocks and knowing names. He is a politician of deceit and lies. And, if you think you are smarter than him, you are God.

He is the devil.

And, he is in the Bible.

This is what he thinks about that is about him. He is the master of conning people and he was created by the church. This is about how he makes you think this and cons you. He thinks he is an angel but he is also a demon known as lucifer. This is about how he cons people and thinks they are stupid. This is how he does life, that is life itself. This is through science and progress. If this is what you learned, then you are wrong. He is much worse than our thought. He is a giant and a monster.

He is a good looking angel.

This is about how The Antichrist looks that is about him. He is like a normal Republican

Naturalization

Senator type that is attractive and good looking. This is about him, and how he looks, and about how he operates. He is the angel of the Lord pretend, and what he does accomplishes nothing. This is about his personality and attitude. He is the prince of light as the prince of darkness. He is about the light but he is too afraid of the darkness that he hides. This is about light and darkness. This is how he is part of it, just by being there when the world was created.

This is about how the way he acts offends people. This is about the way he acts, that offends people. This is about the way he acts, and how he acts. This is pure violence and hate toward normal people. This is about violence and hate toward weird people. This is about people and how he has hate toward them. He gets his power from the church is what he would claim. Yet, all of this is false if he is a false teacher also. This is about how he has been around forever. This is about his lies and deceit that are about the world. The world believes in him, because he is about the world. This is about the world and how he cons it. This is about the world and how he is in it. This is about his false ways and how he acts toward them. He is a angel or a demon, and what he symbolizes is real. This is how symbols work.

This is about the way he works, and how He works, is through God. This is because He is a false teacher. This is about the way he works, that is about the way he works. This is about pure disaster and violence. He is toward all people as good, and what he represents, is pure evil. He is about disaster, and when disaster strikes, he is about pure joy. This is about the world and how it hates The Antichrist. But, we still do not understand what The Antichrist is, which is why he is like a God. He works and operates on evil only, and what He does, is about evil. This is about the overcoming of God, even though he likes him. This is in an angry, messed up way. This is about God, and how he is about Him, is about how evil loves a child. This is in a twisted, messed up, tormented way. This is about the logic of loving. He does not have that, but if he is a teacher, he might. This is about pure falsehood. This is about the logic of the world, and how he becomes it. This is through pure evil. He is about pure absolute power, and how it does not corrupt absolutely. This is about a false teacher's story and how it is corrupt. This is about absolute power and how it does not corrupt absolutely. It is all lies.

The Antichrist has a personality. This is about the corruption of the world, and how it represents itself. This is about hate and violence and how it is corruption. This is about how corruption and hate always do bad. This is toward the world. This is in the eyes of God. This is in the eyes of The Antichrist's personality. This is what the world knows that is corrupt. This is about The Antichrist's personality and how it is corruption. This is about the lies of the world and how they are better than him. This is because he is a false teacher known as the Serpent. This is about his attitude and personality and how it is evil. He is the axis of evil. This is about evil in the world. This is how the world and will represent itself. This is through pure lies.

About the world and about itself, is about what the world is about, and how it is for itself. This is about the format of itself. It is about

Naturalization

religion. This is what rules the world. The Antichrist was created by religion.

This is how the Bibles were created. This is in my knowledge only. This is what knowledge is. It is what we see the world as. This is because it came from the eyes of God. This is how we see the world. This is through the knowledge of good and evil.

How man fell is a mystery. We are not film students seeing everything the actors and acting did. This is about how the world is not knowledge and how the world works is not just knowledge. It is about our life. Yet, I do like to teach the world what I know.

This is about teaching and knowledge. This is about the way the world teaches about the way the world is about the way of the world. This is about the way of the world, and how it becomes it. This is about the way of the world. This is about the way of the world and how we are in it. This is about the way of the world. This is about how we are in it. This is about a system of philosophy. This is about a way of knowledge. This is about the way of knowing and how we become with it. This is about the way of reality and how it has knowledge. This is about the reality of it.

The Antichrist is not it. He is about pure anarchy and destruction. He is about the way of the evil world and how to control it.

This is about the way of the world and how he is in it. These are all devices of the system. These do not make sense. This is how The Antichrist deceives us. This is all lies and deceit. This about the truth and lies.

Naturalization

The Antichrist is the powerful man of the world. He is about all good things. This is about all evil. This is because he changes it. He changes it by walking the earth. This is about his attitude and personality. This is about how he walks the earth and how he does it. This is through communication. He communicates with the earth, through lies and deceit. This is because he is an animal known as the serpent.

This is because he lies like this.

This is about his lies and deceit and how they are powerful.

They are about mind control and about destruction.

This is about the sneaky ways of the world and about deceit.

This is about the way of the world and how it communicates.

This is through the serpent. But, he is the trickiest of all animals. He did tempt Adam to fall. This is from paradise.

This is about the Serpent and its power. This is to communicate to all animals by walking the earth. This is about lies and deceit. This is about a power that deceives the earth. This is about the power that communicates with the earth, and how it communicates with people, is how the snake communicates with the grass. This is about how The Antichrist gets along with the earth. This is about power and corruption.

This is until The Antichrist says he is a humble servant known as Jesus Christ.

Power and Corruption

Naturalization

This is about the anatomy of Jesus Christ. He is about God and about man. He is about the way knowledge works that is about subjectivity. This is about the earth and how it conflicts with man. This is about Jesus Christ as a Carpenter.

This is about good and how it controls us. This is about the world of Jesus Christ, and how He controls them. They are the powerful people who control the world that are about the ones who do control the world. They are about the they.

They are about truth. They have truth that controls us. They are about the elders of the church. They are about the tricky individuals that control us. They are about all the church. They see the world that is ours and thinks that it is theirs. This is because they have supreme knowledge. They are the Ivy's. They are about becoming smarter than you and controlling you. Yet, they are not controversial.

This is because Jesus Christ controls them. He is the supreme judge and smartest man ever. This is about the judge and how it is about him, is about him. He is even the judge of The Antichrist.
The Antichrist gets chained for a thousand years, in the Bible. This is about him. This is because he thinks he is Jesus Christ. But, in all honest opinion, I will not talk about The Antichrist.

Jesus Christ

Jesus is about the power of the world. He is pure good.

Naturalization

Naturalization

This is about the power and corruption of the world. This is about the world and how it has its place with power and control. This is about the world and how it has control. This is over the meek and the powerful. This is about the power and the world that is about corruption and control. This is about the power and control of the world that is about power and control of the world. This is about an earth and a heaven and how to control it. This is through knowledge.

This is through your own knowledge. This is about knowledge and power. This is about the corrupt things in life, and how they are free. This is about Jesus Christ that you must work for. This is about life and liberty and the pursuit of happiness. But, here, I will talk about naturalization.

This is about what I talk about. It is about science and math. It is not about industry. This is about the way of man and how we become. This is about the truth and value of society. This is about the way man talk and walk and how they become. They become through this.

This is about this. This is this. This is about how this is this. This is about how this is this. This is about this that is about this. This is about this. This is this. This is what is this, that is about this. This is this.

This. This.

Naturalization

This is about this. This is about life.

This is about an accurate description of life. This is this. This is this. This is this.

If there were correct answers, this would be the correct answer. This is to this. Nothing controls the world. This is about this. This is about this. This is what this is. Is this about this. This is about this. This is about this. This is what this is. This is. This is this. This is this. The thing that controls us is the thing that controls us. This thing that controls us is the thing that controls all. This things that controls us is the thing that controls them. They are what control us. They are subjective. They change all the time.

Naturalization and Seeing What We Believe

It is about the world that is about the world. It is about the world of the world that controls us. This is about reality of the world and how the world controls us. This is about the collective "we". This is about how naturalization controls how naturalization controls us. This is if we believe. This is how we believe. This is if we believe in Jesus or not. If "we" exists, then we are controlled by it. We are looking for a "one world" answer. This is what "we" get. "But", in reality, it is what we are, that controls us. We are about the realness, we are about.

This is about reality that forms reality. This is the togetherness, that forms what we want, that the connected individuals see. The world is what is confusing. This is if "we" are not a "we" thinker. This is about power and "corruption".

This is how we see it.

Naturalization

This is eventually, if we acknowledge this. This is about the authority and individuals who think this. This is about the whole power of the collected individual. All we root for in philosophy is the free will. All we root for in science is the Nobel prize. We are about togetherness and fun. This is about a world that is together. Together marks the fun we do. This is about the world and how the world functions. This is with togetherness. The story of life book is about the control of man. This is about how the mastermind of myself thinks. Up to this point, I have waited my whole life, just to find out the deeper meaning of life, in a nutshell. This was found by me in the manual of living pretend in fiction. But, in reality, we must explain ourselves depending on how much we understand it. This is the way it is for me, and some others. This is categorized as "soul searcher". This is just my category. Others include, smart, intelligent, and crafty beyond reason. They fall into a category called the "great one". This is about what I have learned and my regrets. This is to not be Satanic and be Christian. This is from my soul searching. This is just a category of knowledge. I did not fall into any categories. I might call myself "most intelligent" as a liar. This is why I think in terms of good and bad. I was raised that way. After writing 49 book, I have considered myself as too intelligent for words. This is like a voice spoke to me. It told me I was the most intelligent. Yet, it was like a lie. This is because I am more intelligent than this. This is just as a writer. As a social person, I am a smartest man ever kind of person, who doesn't like social events, and doesn't like talking as much. But, I am very

Naturalization

smart. This is why I wrote this down. This is to prove The Antichrist wrong. People have more value than him.

This is about people like Charles Leary and Phillip Prestwood. They are in the military. They are highly liked because of this. Yet, most see that military is about good. This is about bad and good. This is about good guy and bad guy. I have learned that most people are good. This is even because of the military. This is how I have learned that most are good. This is about class. I still think in terms of my graduating class in the 2000.

This was Highland Park High School, where I went to go there to graduate. I did do well. But, I had ADD. This is what a real diagnosis was. This was from like it was from my high school. I did not know what shapes the material universe, but I wanted to find out. This is what writing was like.

This is what my real life goal is. This is to really see the unfolding of the universe. This is always the biggest goal in writing. But, for me, to find out what the soul is, is the biggest answer in life for me. But, also, to find out what the brain is, is the biggest answer in life for me. But, this is in transition to what I truly wants, this is to find out what the world means to God, which is a combination of all three.

This is for what it is about and what it is worth. I used to get bored in school, but I became much more intelligent than what I had wanted. I do not want what I wanted in the past in the long run. This is why I do not know why I choose

Naturalization

what I do. Belief in God is what most writing is about, and what most writing is about is about what you believe in. This is the whole science behind what belief is. This is about the whole of belief, and what it does to you. This is about the reality behind what love is about, between a man and a woman. This is about discovering God, and how He becomes. This is in terms.

1. Morality
2. Gradually understanding
3. Seeing Things
4. Believing

This is about how God and how I try to understand the deeper meaning of God through all things in my life. This is about how the world unfolds and how to find a greater meaning of life. This is always "in" Christ, and Christian values. Yet, some of us are meant to explore. This is about how the world unfolds, and the deeper expression of it, is about. It is about it. Find the deeper meaning of life, and how it becomes. This is about the driving force behind social relations. This is about the dependency, of values, and how we treat one another. This is about what the naturalization is about that is about the naturalization of mankind is about that is about what the naturalization of mankind is about. This is about the naturalization of mankind. This is how mankind is about naturalization and how he explains himself. This is to the world.

Man

Man and mankind are about mankind and how he understands the world. This is through

Naturalization

naturalization and the understanding of the world. This is through naturalization. This is about how naturalization understands the world, and how naturalization survives. This is through naturalization and the universe. This is how it unfolds. This is about how the universe unfolds and how it comes back together. This is how the meaning of the symbol of The Antichrist is meaningless. This is about understanding yourself, then going from there. This is about the meaninglessness of life and how we understand it. It is about itself, that is about itself. This is about itself. This is what is about this, that is about this, that is about this. This is about this that is about this. This is about this, that is about this, that is about this. This is about this. This is about what is about this, that is about this. This is about this. This is about this, that is about this. This is about the way the universe unfolds. This is about the way the universe is. It is about this, that is about this, that is about this. This is about this. This is about this, that is about this. This is about this, that is about this. This is about this, that is about this. This is about this, that is about this. This is about this, that is about this. This is about this, that is about this. This is about this, that is about this. This is about how this is about this, and this is about that. This is about this, and that. This is about the world, and how it unfolds. This is about this, that is about that, that is about that. This is about this, and about that. This is about this, and about that. This is about this, and about that. This is about that, that is about that, that is about that. That is about that, that is about that. A fact, is about, a fact, that is about, a fact. This is about a fact, that is a fact. This is a fact, that is about a fact, that is about a fact. This is about a fact. This is a fact, that is about a fact, that is about a fact. That is a fact, that is about a fact, that is about a fact. That is that, that is about the factual existence.

This is about the factual existence, that is about a factual existence, that is about a fact, that is about a fact. This is all factual.

Factual Existence.
Factual Existence.
Factual Existence.
Factual Existence.
Factual Existence.
Factual Existence.
Factual Existence.
Factual Existence.
Factual Existence.
Factual Existence.

All things are possible, through Christ who strengthens me.

This is existence. This is about a fact, that is about a fact, that is about a fact. This is about a fact, that is about a fact, that is about a fact. This is what factual existence, is, that is about factual existence. This is about factual existence. This fact, is about existence, that is about existence, that is about existence, that is about existence, that is about existential existence, that is about existence, that is about existence, in which is about existential existence, that is about existential existence, that is about factual existence, that is about factual existence, that is about factual existence, that is about factual existence, that is about factual existence, that is about factual existence, that is about factual existence, that is about factual existence, that is about facts. This is about all facts. This is about

Naturalization

facts, that are about facts, that are about factual existences, that are about factual existence, that is about factual existence, that is about factual existences, that is about factual existences, that is about factual existences, that is about fact and fiction and how the world is about factuality. This is about factual existences, that are about factual existences, that is about factual existences, that is about factual existences, that is about factual existences. This is about factual existences, and the way they work. This is solely about factual existences. This is about factual existences, that are about factual existences, that are about factual existences, that are about factual existences, that are about factual existences. This is about factual existences that are about factual existences that are about factual existences, that are about factual existences, that is about factual, and representation in which is about factual, and representation, that is about factual existence and the way we treat it, in which is about factual representation, that is about factual existence, that is about factual existences, that is about factual existences. This is what the ways of the world are like. This is about the way of the world and how factual existence is about the way of the world. This is about factual existence, and how the world works with factual existence, and how the way of the world works with factual existence. This is about factual existence, and how the world works through Jesus. This is about Jesus and how the world works with Jesus. This is about Jesus and how He went through factual existence. This is to create all of it.

Before Jesus, there were no bad things. There are no bad things still. This is because

Naturalization

factual existence is for and about Jesus that is for
and about the world that is still about factual
existence. This is about how He changed the
definition of the world to factual existence. This
is about He did all things through Christ who
strengthened Him. This is about what factual
existence, is about. This is about the coming of
Christ, and His standards. It is about Christ, and
what His standards are, are about the living word.
This is about His gentle and focused demure. It is
about factual existence, that is about factual
existence, that is about his demure and focus.
This is like Him, to create this. This is about how
He had focus and demure. This is about His
nature, and what became of man. This is about
how His nature felt, and how we feel. This is how
His earthly self, became man, and rose again.
That is how He became, man, and rose again.
This is about man, and how his self, rose again.
This is how man rose again. This is through
Jesus. This is about how man rose again, because
of Jesus. This is how man rose again. This is
through Jesus Christ. He is about God, and God
is about Jesus, because of Christ. This is about
how Jesus rose, and what He rose, to get there.
This is to Heaven's above. This is about Heaven,
where Jesus rose. This is about how Jesus, rose
again, and how He treated everyone. He was a
modest and humble, servant, of God. This is what
He did, and how He did it. This was in the name
of God. This is about God, and how He found
me. This is about God, and how He found me.
This was it. This was at Young Life camp. He
found me there thrice. This was where He found
me. This was at Young Life camp. This was
where God found me. This was where He found

Naturalization

me, and how He found me. This is when I was confused. This is about what Christ was like, and how He found me was when I was very lost. This was how I was lost, and how I was confused. This is about how I was lost, and confused. This is how I was lost, and confused. This is about how, I was lost, and confused. This is about how I was confused, and lost. This is about how I was lost, and confused about Christ, and His secret identity. This is about how I was lost, and confused, and how I didn't know this, until I was saved. This is why most people act the way they did and do and did and do and did and do and did and do and did and do and did and do and did and do and did in a long period of time. This is until Christ saves you. This is about how the life of Young Life is about. This is about how salvation is. This is at the foot of Jesus. He is also about saving your own life. This is about how friends and fellow campers were also at the foot of the cross, forever and always. This is about the love and compassion of Jesus Christ. This is about the world, and the ways in which it works. This is in very mysterious ways. This is about saving your own life and how it is good. This is about the reality of myself and how He causes all campers to lead themselves to Christian values and Jesus Christ. That is about what is about that which is about that and what is about that which is about what is about that which is about that which is about what is about that. This is.

The "this" which I am talking about just is about that which is about that which is just what it is, and no other thing, other than just this. This is about this.

We will always see the glory of Christ, in

Naturalization

others. This is what separates us from Christ,
unless we become Christians. That is about that
which is about that which is about that which is
about that which is about that which is about that
which is about that which is about that which is
about that. This is about that which is about that
which is about that which is about that which is
about that which is about this and that which is
about that in which is about shape and form. This
is about that which is about that. This is about
that which is about that, which is about that,
which is about that. This is that which is that
which is that which is about that which is about
that. That which is about that, is about Christ.
That which is about that, in which is about this, is
about that. This is that which is about that which
is about that which is about that which is about
that. That is about that, which is about that which
is about that, which is about this that is about that
in which is about this in which is about this. This
is what that which is about that which is about this
that is what that is about that is about what that is
about. This is what this is about. That is what
that is about. This is about what that is that is
this. This is about that. That is what that is about
that is about what that is about. This is about this.
This is about salvation. This is what the deeper
meaning of life is, that drives us. This is about
campers and people like Michael Lavery who are
about and for and about and for. These were the
people who I interacted with, who created me as a
Christian. This is about what Christ is about that
is about Christ. He led me to Christ. Jesus Christ
led me to Christ. This was this that happened.
This is about this that was about this that was
about this that was about this that was about this

Naturalization

that was about this. This was about this. This is about the book that is about the book. This is about what the book is about. This is about what this is. This is about this. This is about this that the book is about. This is about this. This is about what this is about. This is about what this is. This is about what this is about that is about this. This is about this.

War and Peace
This is about war and peace. The extremes of war and peace are very complex and difficult. This is about how war and peace are similar. They are studied together.

But, this is in complex situations. This is about complex situations known as war and peace. This is about war and peace and how they become together. This is about war and peace and how they become together. This is about the extremes of war and peace and how they are naturally together. This is about how war and peace are together.

In an abstract world, there is war and peace. This is only abstractly though. The real reality is though that there is war and peace. There is a combination of both. This is about naturalization and how it doesn't see this. This is about how war and peace must be together. This is about war and peace and how they seem like they are together. This is about war and peace, and how they must be together. That is what war and peace must be.

They are slaves to each other. They are monsters to each other. This is about how different wars, have different themes, and how they represent each other. This is about how there

Naturalization

is themes of war, and how they become. This is about how if there were no war, there would be no peace. There is a theme about every war, and every peace. They are usually the same. This is all ended through the same order. This is what peace and war are created from. This is from knowledge. This is how the dichotomy of knowledge creates war by having peace. There is a theme to everything, so how this relates to war, is similar in how it relates to peace. This is about how becoming, is similar. This is about becoming, to each other, with war and peace. They are together, so they are separate. There is no war and peace. This is without war. But, knowledge says, there is, two sides to every different. This is about how the war is different then the peace, so there is direct opposite. This is about how peace is different than war. War is started in the name of God. This is about peace, because it is its opposite. So, the reality of war is true.

There is a way of becoming and, this way to the other because, war and peace are similar. They are both about victory and loss. This is about war and peace and how they are together about victory and loss. This is about the ways of war and peace and how they are together. Peace is a fiction.

They are about how they are together, and how they are together, are about how they are together. This is about the similarities of one another, and how they are together and both the same. Peace is a lie also.

This is about how they are together, and how they can be separated with history. This is about how history is a lie and it is fed to you.

Naturalization

This is about how history is about how history is about the story. This is merely a fiction tale about gain and loss. The fiction of the world is true. This is about the world being told to you that is about the loss and gain of the world. This is about history that is true, and is told as lies. This is because the bureaucrats of Washington, sit and stare at war, until it comes true, and even go to church about it. I am a writer who thinks differently. This is about peace, and peace. There is no war if there is mind control and violence. This is what I have experienced in my life. This is about how war and peace relate, and how they relate is a strange one. This is about my concentration in life, which is about peace only. This is about peace and war and how they separate. This is about Washington, D. C. and how it doesn't. Peace can be had at D. C. and what peace is, is about world change. This is about reality and how we see it. This is about mind control and how they lie to us. There is no united government or we would all be mind controlled. This is by one single group. This is by the church.

This is why church and state are separate. This is about naturalization and how it is only in schools. This is not about the way it works. This is about the way it is different. It is the same theory of creating God as evolution but it is its opposite.

This has nothing to do with Charles Darwin. The opposite of naturalization is evolution. This is the opposite of naturalization. This is about what state is. This is about direct and total opposites. This is about the strange and peculiar relationship between war and peace.

Naturalization

Similar speaking, the contents of naturalization and evolution are about the similarities between war and peace. If you think you can, you might be able to see them. This is about how war and peace, are similar, if not just equal. This is about the limelight and how you are in it, if you discover this. It will take a mind opener known as Jesus Christ to let you understand it. This is about the war of the world and how peace has it. This is a reality of war and peace and how we don't understand it. This is because of war and peace and their similarities, about Charles Darwin and Riley Miller and their opposites. This is what the real world of religion understands. It understands how the opposing halves theory was sound. This was from my imagination. This is how all things came to be. This is only in an imagined world.

The test of strength is about war. It is only about peace. There is no endless war.

There is no endless evolution. There are lies and truth to war. Peace is pure. There is endlessness in peace. This is about war.

This is about how peace is about how peace is about how peace is about how peace is about how peace is about how peace is about how peace is about how peace is about how peace is about how peace is about how peace is about how peace is about peace. There is no war in peace. This is the only exception of deformation. This is that there is no war and peace explanation in deformation. That is why it seems so powerful. This is because the logic of peace, is about peace. The logic of war, is about war, that is about war. But, there is no conclusion in deformation except that it is about itself. This is all that it is about. There is no war and peace in deformation. This is

Naturalization

about war and peace and its reason. There is none, which is why war and peace are hard to come by.

This means that knowledge is about the way that knowledge is about the way that knowledge is about the way that knowledge is about the way that knowledge is about the way knowledge becomes. This is about becoming that is knowledge, and becoming is about becoming. This is about becoming. This is about war and peace, and how it doesn't become. This just happens. We do not know why it happens.

And, we do not know what it happens to.

This is because of misunderstanding. This is how we have misunderstanding to each other. This is because of war and peace. This is about how war and peace misrepresent themselves. This is on the purpose of the reason. This is to relate.

This is in a direct opposite. This is the direct opposite of church and reality of state and naturalization. War cannot explain naturalization. This is about naturalization that is about naturalization. This is a reality in which can't be described.

This is in terms of all. This is about deformation. This is why it is all. This is because naturalization is about a science of man, that becomes. This is about becoming and how it is not for anything. This means that it is not about war and peace that are a false theory of knowledge. So, you cannot have war and peace in naturalization. This is about war and peace and how we have it. This is about war and peace and how it is an obligation. This is to control itself. This is that it doesn't want itself. This is just with naturalization terms. It does not exist.

This is about the Bible and how it does exist. This is an explanation.

All science is, is an explanation for something. This is an explanation of our reality. This is all it explains.

Factual Existence.
Factual Existence.
Factual Existence.
Factual Existence.
Factual Existence.
Factual Existence.
Factual Existence.
Factual Existence.
Factual Existence.
Factual Existence.
Factual Existence.
Factual Existence.
Factual Existence.
Factual Existence.
Factual Existence.
Factual Existence.
Factual Existence.
Factual Existence.
Factual Existence.
Factual Existence.
Factual Existence.
Factual Existence.
Factual Existence.
Factual Existence.
Factual Existence.
Factual Existence.
Factual Existence.
Factual Existence.
Factual Existence.
Factual Existence.

Naturalization

Factual Existence.
Factual Existence.

This is about factual existence, that is about factual existence. This is about the reality that moves factual existence. This is only science's. This is about how science is an art, and what art is, is about art. This is about how art and science are about art and science. This is about the world that is about the world. This is about the world. This is about how the world is about how the world is about how the world is about how the world is about how the world is about how the world is about how the world is about how the world is about how the world is about how the world is about how the world is about how the world is. It is about what it is. This is about the world. This is about the world and how it relates. This is to everything. This is about how the world relates to everything that is how the world relates to everything. This is about how the world's will relates to everything. It is about how it relates to everything that is how the world relates to everything. This is about it. This is about this that is about this that is about this that is about this that is about everything. This is about this that is about this that is about this. This is about this. This is what this is about that is about this. This is about this. This is about how things are what they are about, that are about how they are about, which are about how they are about things that are about things. This is about it. This is what this is. This is about the thing that keeps us moving. This is about the world and how we become what we want to become. This is about becoming. This is how becoming is about becoming that is about becoming that is about

Naturalization

becoming that is about becoming that is about
becoming that is about becoming that is about
becoming that is about becoming.

Order.
Order.
Order.
Order.
Order.
Order.
Order.
Order.
Order.
Order.
Order.
Order.

This is about the world, that is about
everyday practice. This is about the logic of man
and how it is about ruling. This is about ruling
that is about ruling that is about ruling that is
about ruling that is about ruling. This is about the
everyday practice of ruling. This is what we
wanted. To be ruled by something or rule
something. This is about how communication
works and tells that you are there. This is about
how the world is about how the world is about
how the world is about how the world is about
how the world is about how the world is about
how the world is about how the world is about
how the world is about how the world is about
how the world is about how the world knows
anything. This is about rulers. They are the ones
who pass the laws, and make up the reality. But,
Jesus Christ is King of Kings, so there is no point
in Him being a ruler on earth, because He is the
ruler of everything. This is about the motto that is
about everything that is about everything that is

Naturalization

about everything, that is about everything. This is about everything that is about everything. This is about how the ruler of everything does it. This is through freedom of the will completely.

This is also His power to judge. This is about Jesus Christ and a personal relationship with Him. This is about Him liking what He is about that is about what He is about and what He is about is about the future. He will judge us in the Garden of Eden. This is probably true. Even if it doesn't look right, doesn't mean it is. This is about not looking good and telling people. This is about the world that is about the world that is about the world. This is about the whole world. This is about what the world is about that is about what the world is about that is about the world.

This is for about the how the world works. This is about through sages and prophets. This is about smart and perceptive people. This is how people see the world. But, the real world has nothing like this. The real world has something to expect itself to become, then they become it. This is about how the world relates to itself, and how the world relates to itself, is through becoming. This is about how the world is about how the world is about how the world is about how the world is about how the world that is about how the world is about how the world is about how the world is.

The world is about the world. This is how the world works. This is through the matrix of design of reality. This is about the world and how it has an order. This is how order works. This is through the design of the matrix. This is about how it is designed and how it is formed. This is through all things through Christ who strengthens

Naturalization

me. It is about how the ruling of the world, is
about the ruling of the world. This is about how
the ruling of the world is the ruling of the world.
This is about obeying Christ and doing what He
wanted. This is about the world that is about the
world that is about the world that is about the
world that is about what the world is about, that is
about the world that is about the world. This is
what is about the world that is about the world.
This is about the world that is about the world.
This is about the world that is about the world.
This is about the world that is about the world that
is about the world that is about the world that is
about the world. This is what this is. It is
freedom and the United States government that
runs the world. This is just what it seems. This is
about the world that is about how we feel and how
we become with the people of the world. This is
about a new creation from the government. This
is about how the world works, and how it works,
is through the world. The world has one central
government in the modern day. This is about how
we cope with life. This is through bills and
papers. This is from the desk of the modern day
Jesus. This is from the desk of the modern day
ruler.

This is about science and deformation.
Everything you know is true. This is about the
world and how you represent it.

This is about your life and how you
become it. This is how the life of the will and the
world represent each other. This is through the
reality of the world. This is about the world that
is about what the world is about that is about the
world. This is about the world itself. The world
itself gets involved with everything. It is about

Naturalization

judging others to see what you are like. It is about going to bed, then getting up early to rise. This is about the world that is about the world that is about the world. This is about the reality of the world. This is how we perceive it. This is with clear eyes.

The vision of the way we see things, is the vision of the world. Yet, the difference between me and you is vast. This is about the world and what the world is about, is about the world. This is about the world.

This is how the world is good and how it becomes. This is about the world which is will and representation. This is about how good becomes.

Forever.
Forever.
Forever.
Forever.

This is about eternity and is about eternity. That is what the world is about that is about what the world is about. It is about the world and what the world is about is about what makes the world tick. This is about how the forever in the world is about sin and gloom. This is about the clearness of the world and how we are accepted into it.

It is about the game, which is a part of the chance. This is about the chance to become, something we haven't become. This is in life. This is about the world that is about the world that is about the world. This is about the world that is about the world. This is how the world is about the world that is about the world. This is about the world in which we live. This is about the world, and how it has will and representation. This is about the self. This is how the self works.

Naturalization

It works through the world that is about the world that is about the world. It is about the world, that is about the world, that is about the world. This is all about the world. This is about how the world works and functions. This is about the world and what it does is about the world. This is what the world is that is about the world. This is about the world. This is what the world is about that is about the world. If you do not understand this, you do not understand the world. This is about the world that is about the world. This is about the world, and how the world is about the world. This is about the world. This is how the world is.

This is about it. In clearer terms, the world is what it is, that is what it is. It is about the world that is about the world. This is about the world and how we represent it. This is about the world and will and how we represent it. This is through possibility.

This is how possibility happens. This is about the world that is about the world that is about the world. This is about the world. This is about the world that is about the world that is about the world that is about the world. This is about the world that is about the world. This is about how the way of the world worked. This is about what is through the world. This is about the becoming of man and how he becomes. This is about the becoming of man, and how he becomes. This is through becoming.

This is about a story of a man. He is Jesus Christ.

"I can do all things through Christ who strengthens me." Philippians 4:13.

Naturalization

This is about a book that is about a book that is about a book. This is about a classic novel. This is about the world, and how we represent it.

This is through the Bible. This is about how the Bible represents itself. This is because it is God's word. This is about the Bible and how its structure works for everyone. This is about God's word. This is about how it is about how it is about how it is about how it is about how it is about itself. This is about how the world has changed.

This is for the better. This is how positive reinforcement works. That is what the world is about that is about the world. This world is about the world. This is about the world that is about the world that is about the world that is about the world. This is about God that is about God. This is about all I have talked about. This is for the world that is about the world, which is about the world. This is about what the world is about that is about the world. This is about the world that is about the world that is about the world that is about the world that is about the world that is about the world.

This is about the human that is about the world. This is how he thinks that is about this. This is about what the good of the Bible is about that is about the good of the Bible. It is about the world and how it represents itself. This is about the world, that is about the world. This is what the world is about that is about the world. This is about what the world is about that is about what the world is about. This is about the obvious.

The Nature and Man Process

Naturalization

This is about the nature and man process.
This is about naturalization. This is about how
naturalization is about naturalization. This is
what naturalization that is about naturalization.
This is about naturalization that is about
naturalization that is about naturalization. This is
about naturalization that is about naturalization.
This is about naturalization that is about
naturalization. This is about a process known as
deformation.

This is about becoming. This is how we
have becoming. This is how becoming works.
This is about becoming and how the world works.
This is about naturalization. This is about how
naturalization works, and how it works is for
naturalization. This is about how naturalization is
about how naturalization is about how
naturalization is about how naturalization is about
how naturalization is about how naturalization is
about how naturalization is about how
naturalization is about nature and man.

This is about how he becomes. In this
book I have been becoming as I have written.
This is about the becoming of man that I have
demonstrated also. This is as I have made this to
be. This is through all. This is what man has
done to himself. He has stranded himself and
become with the world. This is about the man
that is about the man that is about himself. This is
about the man and himself and how he is. This is
about the world that is about the world that is
about the world that is about the world. This is
about what the world is about that is about the
world. This is about the world. This is about
what God wants, and what He wants, is about
God. This is about the world that is about the

Naturalization

world. This is about the world that does this to all. This is about the world that does this to all. This is about the relationship of man and his nature. This is about how man and nature are about each other. This is about a good thing that has always been around. This is about always being around. This is how the man becomes. This is with himself. This is what himself is about, and what he is about is about himself. This is about himself, and what he is, is about himself. This is about his environment and his nature verses nurture chronicles. This means that he is about the world. This is what the world is about. This is about the world. This is how the world evolves. This is through evolution.

This is not a science, but a realization that makes you think this way. This is about the way that man verses environment is about man verses nothing. There are no man verses anything in naturalization. This is how his environment is about his nature. All things in naturalization must add up to one. This is all the elements in naturalization. This is about naturalization that is about naturalization. This is about naturalization. This is about naturalization that is about naturalization. This is about naturalization that is about naturalization. This is about naturalization. This is what naturalization really is. It is a creation of the world. This is about the creation of the world, which is about the creation of the world. This is about the creation of the world, which is about the creation of the world. This is about the creation of the word, which is about the creation of the world that is about becoming one with God. It is about naturalization. This is about the creation of the world, and how to become it.

Naturalization

It is about naturalization, which is about
naturalization that is about God. And, this is in
God alone. This is about what man is about that
is about what man is about. This is about man,
which is about what man is about that is about
what man is about. This is about man. This is
how man survives. This is about man. This is
how he survives. This is through the
communication of the world that is about the
communication of the world. This is about the
communication of the world that is about man.
This is about how man conquered the world. This
is not about survival of the species. This is why
there is no bad thing about naturalization. This is
about good and how it does it. This is about how
good is about good that is about good. This is
about good, that is about good. This is about how
the world is about the world. This is about how
the world is about the world. This is what the
world is about. This is about the world and how
the world is about the world. This is about the
world that is about itself. This is about the world
that is about how the world is about itself. This is
about naturalization. This is about naturalization
that is about naturalization. This is how
naturalization thinks and how naturalization acts.
It is a breathing-machine.

This is how fairy tales are made. This is
about finding your identity and your source. This
is how the world unfolds. This is about the
uncovered mystery of the world. This is how the
mystery of the world unfolds. This is about the
mystery of the universe and how it unfolds. This
is about the uncovering of the universe and
unfolding of mystery. This is about how the
universe unfolds. This is about solid expectations

Naturalization

and solid reputations. The fact is, that deformation rules the world. This is just one of the theories that does. It is not about this or not about that. But, what counts is eternal.

This is your relationship with Jesus Christ. This is how you are a newborn Christian. This is only if you wanted to. Naturalization is about the test of strength of mankind and how he operates. This is through the world. This is about the world. This is all about the world. This is how the world is also about itself, and about itself. This is about how deformation looks. It looks like there is a good thing about it. It is about a theory of explanations. These are all about good things. This is about good things. This is about good things. This is about good things. This is what good things are about, that are about good things. This is about good things. This is always about good things. This is about good things that are about good things. This is about the things that are good that are the things that are good. These are about the things that are good. This is about the things that are good, that are good. This is about the things that are good. This is about the things that are good. These are about all the things that are good. This is about good that is about good. This is about good that is about good. This all is about good. This is what good is about. This is about good. These are what the good things that are good are good. These are about good things. This is what is good. This is about good. These are about good. These are about good. This is about pure good. This is about purity and goodness. This is about the value system of our system. This is about nothing much but fun, and nothing much but just plain old values that we learn. This is about values that are about fun, and nothing but fun. This is for the system. This is about the world that is about our world that are about our world that are about our world. These are about the value system of Jesus. This is about values. This is one creation. Another, creation of the modern day is about justice. This is about justice for all. This is why justice is about justice. This is because it is about justice. This is what justice is, that is about justice. This is a higher calling. This is about our system and the design of it, and how it is good with us. This is about how the world is about. The reality of it is, that we are all in the world. This is a world with different situations. This is about the relativity of situations. This is all about the relativity of situations. This is how we learn. These are incidents of modern fun. This is also about violence and sex. This is about how situations are fun. They are also bad. These situations are what the world is about. This is about the becoming of man. This is about the becoming of man. This is how he becomes when he is in his environment. Environment is always around. This is a truth

Naturalization

from the reality of all. This is because of my mind that thinks this. This is about your mind that thinks that. This is about how man and his environment are his nature. This is about how nature is superior to man. This is how the world works, that is about the world. This is about how the world works, that is about environment. We just think it works because of environment. This is about how environmental is how man functions. This is how he is shaped. This is also true of Darwin and Freud; they were shaped by their environment. This is about how man and nature are about how he is really shaped. This is through the environment. This is how the environment shaped man. This is about how man shaped his environment.

One example of how the world was created. Diagram 1.

Creation of the world, one theory in naturalization guided by context.

In terms of naturalization. This complex equation is an answer to the riddle of how, once the earth was created, how it came into is. This is part of naturalization.

It could be an answer to how to explain both, in different ways of communicating with parts, could explain the change in environment theory. There are aspects of both big and small ways of communicating within limits. These limits are set in the equation world.

This change in environment theories has aspects of both design and both limits of what we could be. This is in nature, in man, in environment, and in turn in naturalization. It is

Naturalization

one of the most challenging aspects of God out there.

$A + B = C = C$ (where the second C is a + b also). In a relationship of part to part, and one whole, the second part is two parts and one whole both. This is then much more complex and divided into much more complex difficult equations. This is one whole equation. The words in between describe the parts necessary to get to the ending.

Naturalization is this process. The process of undermining naturalization.

C is the product of the first problem, is A + B equaling rightly C.

The second answer to the problem is C, from A + B. This is from part to part, to one whole.

The real C, is the right answer.
To get there, you must complete these steps.

It should look like this in your mind. Something like: $< A + B = C > = < A + B >$ Or $< A + B > = C$ Or, in example A plus B equals the equivalent of A plus B equal C.

The equivalent is A + B. In exactitude, it would look like $A + B = C = C$ where the end C is A + B.

I. The question of naturalization.

Problems:
This is what you get after seeing what these steps can be, in diagram 1. (see diagram 1).

Diagram 2.

Naturalization

$A + B = A + B + C$. A plus B equals the equivalent of $C + C$. So the end equivalent to the first equation is C. This is after $C + C$ is equal to C. C is equal to $A + B$ in the first equation, so C is equal to $A + B$ in the second equation. The second equation is in diagram 2. These are representative. These are equal in equation, but less equal in the existent equation of exactitude of the number. The real value is a number that is far too complex to understand.

This is why it is written in words. Words have meaning beyond the value of letters. This is the same as numbers. This is from the top, where it says "This is in nature. This is in naturalization. "This is in nature. It goes on and continues to a long while. This is what naturalization is; it is an example of what goes into being on top. This is what equations are. This is why it is confusing, because it is like an equation in itself. This is why it is confusing.

Because it is naturalization.

This will be figured, from the problem, in that of. That of which is occupied in time and space and thinking.

This is the question of the difficulty.

This is the right start.

The first order signifies the naturally inclining nature of man to direct himself toward an order, of man, in nature, that is his environment. This is the process. This is of naturalization.

This is simple.

This is what solves C at the end.

It is needed to know what the problem is, before you do it. The C is the final answer, so the problem here is the process to get to the answer. This is the thought involved in the thought process. Once finding C, then the problem can be solved. This is the naturalization theory of nature. Naturalization is the answer. This is the complex need of society to fulfill itself. A game of society is fulfilled within the problem. It is to solve the problem. This is why society is superior. The whole of this problem is an answer. All of this equals to one conclusion. This (up above) should be seen as a whole.

= answer (C) here

Diagram 3.
$C = A + B = C$

Last part is factored in with the second and third parts equal due to a thought process Figure 1. This equals the answer.

$$A1 + B1 = C1 \text{ Whole problem}$$

First Part: A

The order of numbers in the problems, are orders that follow the guide. This is the real naturalization processes of nature. It is a cause and effect relationships of cause and effects. Therefore, the numbers increase as you go onward toward .

$$(\quad A1 + B1 + A2 / C1 =$$
$$A3 + B2 + A4 / C2$$
$$= C2$$

Naturalization

+

A1 + B2 + A2 / C1 =
A3 + B3 + A4 / C2

= C2

+

A1 + B2 + A2 / C1 =
A3 + B3 + A4 / C2

= C2

)

(+)

Second Part: B

Proofs

(A1 + B1 + A2 /

C1 =

A3 + B2 + A4 /

C2
= C3

Proofs

A5 + B3 + A6 /

C4 =

A7 + B4 + A8 /

C5

= C6

Proofs

A9 + B5 + A10 /

C7 =

A11 + B6 + A12

/ C8

=C9

)

Naturalization

First part A, and second part B, are the same.

Numbers here are symbolic of every different number that is in the problem. These numbers represent a good look at the number system. This is true for all of these problems.

They just have different scaling numbers. The numbers on the scale are for different purposes. This one is to show what it can happen to have when numbers go bigger. The bigger consequences are bigger on the second scale. Then naturalization, one environment, one nature, at this point, is a better bet. A naturalization process of nature and man are congruent with naturalization and environment and man. These results show in the second form. This is because the numbers get bigger, which means it is the same, bigger is better. This is because naturalization is the same. But, to prove it, we must go through all the steps. Man + environment = nature = naturalization. This would be too small. This needs to be expanded. The need to expand is needed in this expansion. This is part of naturalization. The problem is in itself, so the expansion on thought would require separate thought process. The result is nature. This is the natural result. This natural result is filled with the natural. This is the naturalization of man.

This is in problems.

The answers are plugged in. This is in the second part.

But man isn't part of nature. Man is part of naturalization.

First, we understand nature.

Naturalization

The proofs of this show the end result. This is at bottom after the word or. This is after putting this problem into the thought process.

The thought process translates to:
(Thought process) Figure 1

$$
\begin{array}{llll}
(& & A+ & A + B = A + \\
& B = C & &) \\
& & B & A + B = A + \\
& B = C & & \\
& & =C & A + B = A + \\
& B = C & & \\
& & =C & A + B = \\
& C & & \\
& (& & \\
&) & &
\end{array}
$$

Plug part c (which is the into b, with the problem above (figure 1), and result will be equal. Then it will bring you to the last point.

It translates to:
After figuring the thought process into the problem above. Then it figures into the problem below. The thought process is the combining of the proofs with the problem below (which results in the final problem).

$$=$$

Third Part: C
Initial equation: A which is 1 here, + B which is 2 here
= C, which is three here

$$
A (A + B = C) +
$$
$$
B (B + A = C) =
$$

Naturalization

$$C \ (= (A(123456789,10,11,12) + B(123456) =$$
$$C(123456789)) = C$$

$(((1 + 2))$ or $3 = ((3)) +$
$(((1 + 2))$ (or equaling 3) $= ((3)) =$
$(((1 + 2))$ (or three) $= ((3)) = ($ three $)))$
Remainder is plus the remainder of all problems,
in the remainder of the situation, which three
equals the remainder of all three equations of
math in this context where there is the three that
equal the three at the end. This is an infinity
equals infinity. This is how naturalization or the
circumstance of happening circumstances of
knowledge, work their way into the picture.

Final Answer:
This is how God created the earth.
Answer
Where it equals C.
The third part goes into the answer. This is the
problem to find the answer.
A: $A1 + B1 + C1 = A1 + B1 + C1 = C2 +$
B: $A2 + B2 + C2 = A2 + B2 + C2 = C3 =$
C: $A3 + B3 + C1 = A3 + B3 + C2 = C4$
$= C$
Numbers

After factoring the second and third parts
equally, one comes to terms with the second part
of this problem, which is what it seems. This is
actually one of the last steps in the entire problem,
but it is part of the final answer.

$\underline{(A + B = C)} = C$

Naturalization

Where a + b = c + a + b = C all equal to C. (of all the numbers) of the containing a and b and c, like a1, b1, c1.

It is all combined to equal a thought. This is why it is all equal to a variable, in the end. This variable stands for naturalization.

It is a process, so the conclusion is naturalization.

Translated: powers go to

And C is the final product for all the naturalization and everything else.

$$(A + B = C) \rightarrow C$$

Answer: C is the final answer to all of this

From up above (A + B = C) = C

The last one is the proofs together (from above) with the thought processes. This is the thought process after it worked from up above, changing the variables to make sense. The fact that this was the thought process is a fact of God. This is in the nature of the problem. This fact of God, in nature, is the naturalization of man. Yet, man is not God, so it cannot be solved completely. This is naturalization in man. This is a thought process from man. This is the process of thought for the proof of naturalization. The proof of naturalization is the end result. This proof is naturalization, in order to appeal to the senses. This is the real result, after figuring out how to make the thought process work in the

Naturalization

problems. The second and third problem. The first is added to the second and third. The thought process makes it like there is only one.

Where $A + B = C$ where each a and b and c have each (sub one, sub two, sub three with each word that it attaches to that is), part factored eventually. This is of the second $(A + B = C)$ part. (The one on the right). The one on the right is factored evenly. This is from vice versa factoring, where one factor is on the right and one on the left and one on the far right. This breaks down into a complex problem, of exchanging variables for one another, with each variable on different sides and in the middle.

It is a complex way of congruent math with the first whole part. It is the same as the first part, but with the first part the same, but different, because it is after being solved by the others. Therefore, it is a gradual incline to the answer, rather than a mixing and matching way, which in the end, is, (the same as up at top or similar), but also is more organized. It is an absolutely conjugate way of dispersing the variables in the way they are organized up top. This is part of the problem. It gradually reaches an answer.

An problem that is like this is a fact. This is a fact that is like this. The fact, that it is restricted, is a fact of access. This is a fact that is limiting its ability. The right fact, at the right time, is a correct fact. This is a fact, which is correct. This fact, that is correct, is a fact of a think. This think is correct. This is a fact. This is a fact that is a think. The think is a fact, that the think is correct. This is correct.

Naturalization

II.
Steps from problem:

A.
Step 1:

A1 nature + B1 man + A2 environment =
/ C1 nature =
The nature of man in his environment,
take away his nature equals at first step of three,
merely his environment, which in the next
problem equals

Step 2:

A3 naturalization + B2 man + A4 environment /
C2 naturalization
= C2 natures

…the naturalization of man in nature, take
away his naturalization (from one), would equal
this in steps three and four of his final problem,
yet it doesn't make sense.
This is yet to come. His naturalization
takes place here. To come to terms with
naturalization, one must first recognize the man in
nature which is his environment.
This natural phenomenon of, nature of
naturalization in his environment, in specific, his
nature,

Step 3:

Plus the naturalization of man in his
nature, take away man…

Naturalization

Answer to this (plus)
A1 natures + B2 naturalization + A2 man =
/ C1 man
The man in his naturalization then man
would take away man.

This would then cause the additional next
steps to happen.
They start by adding the nature of man to
his environment, then the natural process of man,
which would mean, in turn that nature. These are
naturalization.

Step 1:
Proofs
A1 + B1 + A2 / C1 =
A3 + B2 + A4 / C2
= C3
Environment + naturalization + nature /
naturalization =
Naturalization + environment + naturalization/
nature =
Man
Step 2:
Proofs
A5 + B3 + A6 / C4 =
A7 + B4 + A8 / C5
= C6

Nature + naturalization + man / naturalization =
Naturalization + nature + man / environment =
Nature
Step 3:
Proofs

Naturalization

$$A9 + B5 + A10 / C7 =$$
$$A11 + B6 + A12 / C8$$
$$=C9$$

Man + nature + environment / nature =
Man + environment + nature / man
/ Nature

Plus symbol is relative to adding the words or numbers together, to get an answer. The relative natures of man to his environment, to conceive of naturalization, would be to divide nature over environment and man, adding nature. This would divide nature over nature in this equation. And then environment over nature, in this same process, after adding man to nature, and dividing man. Man + nature + environment / nature = man + environment + nature / man and then over nature. Your final product always in these equations is naturalization. The right equation to solve is in the first and last endings in a vex.

In this equation, at the end of the book, man plus nature would equal environment. This is the final equation's proof.

Next would be to add the product of the separate entities together, to solve the complex problem of naturalization and the environment, in nature and man.

The man in nature, is man in mankind, which equals the nature of man in mankind. This is in naturalization.

B. The logic of mankind.

Step 1:

Naturalization

Nature + naturalization + man = nature + naturalization + man = man

The man in his environment is natural.
The nature of man in his environment is nature.
This is naturalization.
This is from nature, to nature, to natures.

Step 2:

Man + environment + naturalization = man + environment + naturalization = Naturalization

The man in his natural environment is a man in his natural environment.
This is a product of naturalization.

Step 3:

Man + nature + environment = man + nature + environment = naturalization

This will explains the final product in terms of the right answer.

B.

Final Thought process:
After factoring in these problems.

Naturalization

Part C.
Where it equals C.

$A + B = C$

  After factoring the second and third parts equally, one comes to terms with the second part of this problem, which is what it seems. This is really one of the last steps.

$(A + B = C) + (B + A = C) =$
C
(C is this below).

$\underline{(A + B = C)} = A + B = C + A + B = C + A + B =$
$C + A + B = C + A + B = C + A + B = C$
Where a + b = c is a + b = c (of all the numbers)
of a containing a and b and c,

Translated: powers go to

_____

And A is the final product for all the naturalization and everything else.
Then it equals C.
$\underline{(A + B = C)} \rightarrow \ C$

Answer: C is the final answer to all of this

From up above $A + B = C = C$ The right diagram.

  The final conclusion to this is basically (all of this):

Answer:

Naturalization

Final Step:

The containments of knowledge.

A (Environment + nature + naturalization
= nature + man + environment) +
B (Nature + environment + naturalization = nature
+ environment + naturalization) =
C (Man + nature + naturalization = man + nature
+ naturalization)

B (Naturalization + nature = environment) + A +
(nature + environment = naturalization) + C
(naturalization + nature + environment = nature)
where in this context environment equals nature.
This is naturalization in that it sees that it
is clear. It is seen clear. This is what it is seen.
This is as naturalization. This is what
naturalization is all about, for understanding and
comprehending problems that are stylistic, and
modern.
= Naturalization (C)

This is about how the world is about how
the world is about how the world is about how the
world is about the world. This is about how the
world works. This is through the world. This is
not in and of the world. This is only in
Christianity. This is about the reality of the
world. This is about how the reality of the world
was created. This is how man and nature created
the world. This is the world around you. This is
the environment that I was talking about. It is
also true that environment shapes personality.
This is about the gain and loss of the world. We
must succumb to the world's pressure to become

Naturalization

it. This is a meaning, meaning that there is a higher calling in life, that is beyond reason and rhyme. And, this is naturalization. This is how naturalization works, and how it works, is how it becomes. This is not in factual existence. This is in the mind. This is about how the mind works, and how it works is about fairy tales. These are success stories that are about man and environment. This is about the success of the world, and how environment shapes everything. But, if we didn't have nature, then it wouldn't shape everything. This is about man's nature, that is about man's nature, that is about man's nature, that is about man's nature, which is about man's nature. This is about the man and his nature that shape him. This is all about nature and man's comparison to it. This is about itself, and what is about itself, is about the nature of man. This is about power and corruption. This is just because of knowledge. This is how knowledge works, and how it works is preposterous. This is to all. This is what knowledge is. It is about the way of man. This is how he does his thing. This is how he moves through the world. This is about how we learn it, and how we become. We become like the world, after learning about it. This is about the genius of the world and how it is compared. This is to man and nature. This is about how man and nature do both things to both things. They all do both things to both things. This is the conclusion, that they are all eventually one. This is about how it works, and how it is maintained. This is through science and math, and words. This is about the reality of the world, and how it is changed, to become, what it becomes. This is about the confused reality of naturalization. This

Naturalization

is all about the world of naturalization, and how it worked. This is through the naturalization world of learning. This is about how we learn, and how it is about learning, that it is about learning. This is about learning.

This is how we learn about naturalization. It is a process of naturalization in stores. Take, for example, reading a book you like. This is what the book that is read is like. We are in naturalization the whole time. This is about how pain hurts. This is about after reading too many books. This is about the brain and its anatomy. This is what the world does, that is about the world. This is about the way the brain works, and its anatomy. This is about the brain, and its anatomy. This is about the reality of the world, and how it shapes us. The reality of the world is just environment, which explains this. This is about environment. This is how man and nature shape each other. This is about the nature of man that is the biggest one. This is because man's nature, sees naturalization. This is how it is naturally. This is about nature and naturalization that explain man. This is about nature and man that explain God. This is about God that is about God that is about God. He is about nature and naturalization, because He created man in His own image. This is how man relates to God. He is about the process of naturalization. This is about how naturalization is about God and man. It is about man and God, and how they do it. They do it together. This is in the same way that they live life. They live life together. This is in Heaven. They are the judge.

This is when Jesus judges. This is how Jesus judges that is rightly His. He is about love

Naturalization

and works. This is about how the works of the world are Jesus' and not the Antichrist. He is not in this theory in any way, shape, or form. This is about how the naturalization of the world, is about the naturalization of the world, is about, the naturalization of the world. This is about the naturalization of the world that is about the naturalization of the world. This is about the naturalization of the world that is about the naturalization of the world. This is about the naturalization of the world.

Naturalization is naturalization as naturalization. This is about the world that is about the world that is about the world. This is about the process that is about and what it is about is what is about what is about naturalization. This is about naturalization. This is what the process of naturalization is. It is about and able to be better. This is than other thought. This is about the way naturalization does it, that is about the way naturalization does it. This is about naturalization. This is about naturalization and how it does it. This is about the naturalization that is about others. This is about the way shape and form of reality that is about the calling to duty. This is of naturalization. This is how naturalization shapes the world. All it is, is a theory I created. This theory is about knowledge that is about form, shape, and way. This is about the theory I created. This is about contents of knowledge, and how I got there. I got there through learning about naturalization. This is after I wrote it. This was supposed to get a good grade. But, I was only writing when I did it. This was not in schools. This is about what the grade is about. It is about excellence.

Naturalization

I created naturalization independently
from school. This is why this theory looks
different than in school. This is about our theory
that is in schools, that is in school. This is about
look and design.

It takes the best shape and form of a
triangle with all sides that are the same. This is
what the world offers to itself, which offers to
itself. This is about analysis that is about the
good of mankind.

It is about knowledge that is about
knowledge that is about problem. This is about
the nature of man, and how he works. This is
about man's difficulty with knowledge. It is a
theory of how man became. This is after he fell.
This is difficult to tell, but it is strange. This is
this way, because no one understood how Adam
fell.

This is ever. This is because mankind fell
because of the problem of knowledge. He
couldn't even explain it before, or after he ate the
apple. This is about the apple and the fruit of
knowledge, knowledge problem. This is about
the problem, of science, and math. This is about
the problem of words. They are fully meaningful,
no matter what.

This is about shape and form, which is
about deformation. This is what the matter is
about. This is about explanation.

This is about explanation and how it
explains things. This is to the meek or strong.
This is about learning. This is how one learns.
One learns by being like another. This is about
how we learn, and how learn. This is about how
learning is about learning and about learning.
This is about learning and how learning is about

learning. This is about the prodigals of modern man. This is about his learning and about his becoming. This is about learning that is about learning, that is about learning. This is about learning.

This is how learning is about learning. This is how one thing leads to another that is about learning process and problems. This is totally about school.

Chapter 5
Becoming of Man

This is about man and how he struggled through life. This is about the world and how man is about the world. This is about how man is about the world. This is about how Christ came to save. And, He did save. This is about how man crucified Christ, because of false teachings from bad men. They are not in the Bible. The Bible is God's word, and He becomes with the Bible to know all of it. This is how man is without his Bible. This is about man and his Bible and what he saw. He saw the world becoming to an end. This was about the entire world. This is what the world would be, becoming to an end. This is about what the world is about that is what the world is about. This is about the world and how it does it to you. This is because of the power behind all knowledge. This is the power behind all knowledge. I do not know what it is, but it is very good.

This is named Jesus Christ. This is what Jesus Christ did, that is about how Jesus Christ did. This is about what He did. He was born God. This is about how God did this, and how He did it was through good. This is about how He did this, and this was good. This was about good. This is how He did it, and what He was about, was about good. This was about good things that are about good, which are about good things. This is about good that is about good that is about good. This is about good and how we see good. This is to become and be about good. That good is about that good that is about this good. This good, is about that good, which is about this good. This is about good that is about good. This is about good, that is about good. This is about Heaven, in which is about, good, in which is about how we can be saved. We are about God, because He is in Heaven. And, we want to go there. It is about God, which is about God, which is about God. God is about God, which is about God. God is about Jesus Christ. This was His only Son. This is about God, which is about God. God is about God, which is about God. God is about God, which is about the God of the world. This is about God, and what God is about, is what God is about. This is about God, and what God wanted, was about God. This is about God. This is about God that is about God, that is about God. This is about God. This is about God, which is about God, which is about God. God is about God, which is about God. God is about God, which is about God. This is about what God is about. This is about what God is about, that is about what God is about. This is about good. This is about a real power of God's that is about immortality. This is about seeing into the world and creating it. This is about the world. This is how God created it. This is into a perfect world.

Naturalization

This is about God's power, and how overcoming it, is impossible. No one can.

This is about what God wants. God wants all to become. This is in His image. This is Jesus Christ's. This is about His image that is about what His image is about, that is about what His image is about. This is about His Image that is the best. This is about naturally determining us. This is with Christ. This is how Christ naturally determines us. This is about His will. This is about His will that is about us. This is about us. This is about the way the world is about the way the world is about the way the world is about the way the world is, that is about the way the world works. This is about the way the world works, that is about the way the world works. This is about the way the world works. This is about the world and the way it works. This is about the way it works that is about the way it works. This is about the way the world works. This is about the way the world works. This is about the world and how it works. This is through the world that is about the world that is about the world. This is about the world and how it works. This is about the world and how it works. This is about the world. This is about how the world works that is about the works. This is about how the world works that is about the world. This is about the way it works. This is about the world and how it works. This is about the world and how it works. This is about the world that works. This is about the world. This is what the world is about that is about the world. This is about the world. This is about the world and how it works. This is about how it works and how it works is about it. This is about it. This is what this is about. This is about

Naturalization

the world. This is about the world and how it is
about the world is about the world. This is about
the world. This is about the world that is about
the world that is about the world that is about the
world that is about the world. This is about the
world that is about the world that is about the
world. This is about the world that is about the
world that is about the world. This is about the
world that is about the world. This is about the
world that is about the world. This is about the
world that is about the world. This is about the
world. This is what the world is about that is
about the world. This is about the world that is
about the world. This is about the world that is
about the world that is about the world that is
about the world that is about the world. This is
about the world that is about the world. This is
about the world. This is how the world is about
the world. This is about the world. This is about
the world that is about the world that is about the
world that is about the world that is about the
world. This is about the world and what the
world is about the world that is about the world.
This is about the world. This is about how the
world works, that is about the world. This is
about the world that is about the world. This is
about what the world is about and what it is about
is about the world. This is about the world that is
about the world that is about the world that is
about the world that is about the world that is
about the world that is about the world that is
about the world. This is about the world. This is
about the world that is about the world. This is
about the world that is about the world. This is
about the way the world is about the way the
world is. It is about the world. This is what the

Naturalization

world is about that is about the world. This is
about the world. This is about what the world is
about that is about the world. This is about the
world that is about the world. This is what the
world is about that is about the world. This is
about the world that is about the world. This is
about the world that is about the world that is
about the world. This is about good that is about
good. This is about the good that is in the good in
which is about good. This is about good. This is
about the good that is about the good that is about
the good. This is about the good that is about the
good. This is what the good is about that is about
the good. This is about the good. This is about
the good that is about the good. This is about the
good that is about the good. This is about the
good. This is what the good is about that is about
the good. This is about good that is about good.
This is about good. This is about good that is
about good that is about good that is about good.
This is about good that is about good. This is
about good that is about good. This is about pure
good. This is what good is about that is about
good. This is about the good. This is what good
is about that is about the good. This is about good
that is about good. This is what good is about that
is about great and wonderful things. This is about
what good is about that is what good is about.
This is about good. This is what good is about
that is about good. This is about the good that is
about the good that is about the good. This is
about good that is about good. This is about good
that is about good. This is about good that is
about good. This is about the good of the world.
This is about what good is about that is about
good. This is about good. This is about good that

Naturalization

is about good that is about good that is about
good. This is about good. This is about good.
This is about good that is about good. This is
what good is about that is about good. This is
about good that is about good. This is about good
that is about good. This is what good is about that
is about good that is about good. This is about
good, that is about good. This is about good.
This is what good that is about good. This is
about good. This is what good is about. This is
about what good is about that is about good. This
is about good. This is about good. This is about
good. This is about good. This is about good.
This is about good. This is about good. This is
about good that is about good. This is about
great. This is about good. This is what good is
about that is about good. This is about good.
This is what good is about. This is about good.
This is what Heaven is for. This is about good.
This is what good is about that is about good.
This is what good is about. This is about good,
that is about great and wonderful things. This is
about good that is about good things that are
about good things. This is about the good that is
about the good. This is about the good that is
about the good. This is about the good of the
world that is about good. This is about good.
This is what good is about that is about great and
wonderful things. This is about good that is about
good. This is about good. This is about good that
is about good that is about good. This is about
good that is about good that is about good things
that are about good things. This is about good
things. This is about good things. This is about
good. This is about good things that are about
good things. This is about good things that are

Naturalization

about good things. This is about the good that is
about the good. This is about good things that are
about good things. This is about good that is
about good things. This is about good things that
are about good things. This is about good things
that are about good things. This is about good
things. This is about good things that are about
good things. This is about good things. This is
about good things that are about good things.
This is about good things that are about good
things. This is about the world that is about the
world. This is about the world that is about the
world that is about the world. This is about the
world and what is about the world is about the
world. This is about the world that is about the
world that is about the world that is about the
world. This is about the world that is about the
world that is about the world. This is about the
world. This is what the world is like that is about
the world. This is about the world that is about
the world. This is what the world is like. This is
about the world that is about the world that is
about the world that is about the world that is
about the world. This is about the world that is
about the world that is about the world. This is
about the world that is about the world that is
about the world. This is about the world that is
about the world that is about the world. This is
for and about the world. This is about the world
that is about the world that is about the world.
This is about the world that is about the world that
is about the world that is about the world that is
about the world that is about the world. This is
about the world that is about the world that is
about the world that is about the world. This is
about the world that is about the world that is

Naturalization

about the world. This is about the world that is
about the world. This is about the world that is
about the world that is about the world. This is
about what the world is about that is about the
world. This is what the world is about that is
about what the world is about. It is about the
world. It is about the world and what it is about.
This is about the world. This is what the world is
about that is about the world is about. This is
about the world. This is about the world. This is
about the world that is about the world. This is
about the world that is about the world. This is
about the world that is about the world that is
about the world. This is about what the world is
about that is about the world. This is about the
world that is about the world. This is about the
world that is about the world. This is about the
world that is about the world. This is about the
world that is about the world. This is about the
world that is about the world. This is about the
world that is about the world. This is about the
world that is about the world. This is about the
world that is about that is about the world that is
about the world. This is about the world that is
about the world. This is about the world. This is
what the world is about that is about the world.
This is what the world is about that is about the
world. This is about the world that is about the
world that is about the world. This is about the
world that is about the world. This is about the
world that is about the world that is about the
world. This is about the world that is about the
world. This is about the world that is about good.
This is about that is about the way is about the
way that is about the way that is about the way
that is about the way that is about the way that is

Naturalization

about the way that is about the way that is about
the way that is about the way that is about the way
that is about the way that is about the way that is
about the way that is about the way that is about
this. This is about this. This is how the
interconnected world competes with the world of
individualism, which is about not being
connected. This is to the way the world works
that is about the world that is about the world that
is about the world that is about the world that is
about the world that is about the world that is
about the world that is about the world that is
about the world that is about the world that is
about the world that is about the world that is
about the world that is about the world that is
about the world that is about the world that is
about the world that is about the world that is
about the world that is about the world that is
about the world that is about the world that is
about the world that is about the world that is
about the world that is about the world that is
about the world that is about the world that is
about the world. This is about the world that is
about the world that is about the world. This is
about the world that is about the world that is
about the world that is about the world that is
about the world that is about the world that is
about the world. This is about the world that is
about the world that is about the world. This is
about the world is about the world that is about
the world. This is about the world that is about
the world that is about the world that is about the
world that is about the world that is about the
world. This is about the world. This is about the
world and how you are in it. This is about the
good of the world. This is about the good of the
world. This is about the world that is about the

Naturalization

world that is about the world. This is about the
world that is about the world that is about the
world that is about the world. This is about what
is about the world that is about the world. This is
about the world is about the world that is about
the world is about the world. This is about the
world that is about the world that is about the
world that is about the world. Its is about the
profundities of life that are certain, and the
stupidity sad. This is about the world that is about
the world that is about the world that is about the
world. This is about the world that is about the
world that is about the world that is about the
world. This is what the world is about that is
about that is about the world that is about the
world. This is about the world that is about the
world. This is about the world that is about the
world that is about the world. This is about the
world that is about the world that is about the
world that is about the world. This is about the
world and what it is for and about. This is for and
about the world and what the world is for and
about that is about for and about the world that is
about the world. This is about the world that is
about the world. This is about the world that is
about the world. This is about the world that is
about the world. This is what the world is about
that is about the world. This is what the world is
about that is about the world that is about the
world. This is about the world that is about the
world that is about the world. This is about the
world that is about the world that is about the
world. This is about the world that is about the
world that is about the world. This is about what
the world is about that is about what the world is
about. This is about the world if it is about the

Naturalization

world. This is about the world that is about the
world that is about that is about the world. This is
about the world that is about the world that is
about the world. This is about the world that is
about the world that is about the world. This is
about the world that is about the world. This is
about the world that is about the world. This is
about the world that is about the world. This is
about the world that is about the world. This is
about the world that is about the world. This is
about the world that is about the world. This is
about the world that is about the world. This is
about the world that is about the world. This is
what the world is about. This is about the world
and how it is shaped. This is through thought and
the form of thought that is thinking. This is about
thinking that is based on thought. This is about
thought that is about thought that is about
thinking. This is about thinking that is about
thought that is about thinking. This is about
thinking that is about thought. This is about the
thing that is about the things that are about the
things that are about the thought. This is about
the thought that is about the thought. This is
about what this is about the thought that is about
the thought that is about the thought. This is
about the thought that is about the thought that is
about the thought that is about the thought that is
about the thought that is about the thought that is
about the thought that is about the thought that is
about the thought that is about the thought that is
about the thought. This is about the thought that
is about the thought that is about the thought that
is about the thought that is about the thought that
is about the thought that is about the thinking that
is about the thought that is about the thought that

Naturalization

is about the thought that is about the thought that
is about the thought that is about the thought that
is about the thought that is about the thought that
is about the thought that is about the thought that
is about the thought that is about the thought that
is about the thought that is about the thought that
is about the thought that is about the thought that
is about the thought that is about the thought that
is about the thought that is about the thought that
is about the thought that is about the thought that
is about the thought that is about the thought that
is about the thought in which is about the thought
that is about the thought. This is about the
thought that is among us. This is with the
profundities of wisdom. This is about the way
that wisdom works that is about wisdom. This is
about wisdom and how it works in the way of the
universe that is about the way that the universe
works that is in the way that the universe works.
This is about the way that the world works that is
about the way the world works. This is about the
way the world works that is about wisdom. This
is about the way the world works. This is about
the world works that is about worldly wisdom.
This is about the world that is about the world that
is about the world that is about the world. This is
about the world and how it is about the world.
This is about the world and how it works. This is
about the world that is about the world that is
about the world that is about the world that is
about that is about that in which is about that.
That is about that. This is about that that is about
that. This is about that.

Become Together in Knowledge

Naturalization

This is about the way of the world. This is about the way of the world that is about the way of the world. This is about the way of the world that is about the way of the world that is about the way the world works. This is about the world that is about the world that is about the way the world works. This is about the way of the world that is about the way the world works. This is about the world that is about the way the world that is about the world works. This is about what the world is about that is about the way the world is about the way the world works. This is about the way the world works, that is about the world. This is about the way of the world and how the world works. This is about the way the world works, that is about the world. This is about the world that is about the world, that is about the world. This is about the world. This is about the world that is about the world that is about the world that is about the world that is about the world. This is about the world that is about knowledge only. Writing is the way of the world that is about the way the world unites and celebrates. This is in knowledge. This is what knowledge is about that is about knowledge. This is about knowledge that is about knowledge. This is about the knowledge that is about knowledge, that is about knowledge. This is about knowledge of knowledge that is of knowledge. This is about the way we put it all together. This is about the way we become. This is all about knowledge. This is about knowledge that is about knowledge that is about knowledge. This is about it that is about it. This is about it that is about it. This is what this is about that is about this. It is about what it is about that is what it is about. This is about it. This is what it is

Naturalization

about it that is about it. It is about it. It is about it that is about it that is about it. This is about it. This is what all of it is because. This is what it is about, that is about it. This is about what it is about that is about it. This is about this. This is about this that is about this. This is about what this is about that is about it. It is about this that is about this that is about this that is about this that is about this. This is about this. This is about what this is about that is about this. This is about this. This is this that is that is that. This is about that is that. This is about that is about that is about this. That is about that. That is about that is about that is about that. This is about that. This is about that. That is about that. That is about this. This is about that. That is about that. That is about what that is about that is about that. This is about this. That is about that. This is about that. Which is about that is about that that is about that is about that. This is about that. This is about that that is about that. This is about that. That is about that is about that is about that is that. This is about that. That is about this. This is about that that is about that. This is about that. That is about that that is about that that is about that that is about that that is about that that is about that. This is about that. This is about that that is about that that is about that. This is about that that is about that. This is about that that is about that that is about that. This is about that. This is that that is that is that about what is about what is about that. That is about that. This is about that. This is about this. This is about this that this is about that is about that. That is about that. This is about what this is about that is about what this is about. This is about what this is

Naturalization

about. This is about what this is about that is about what this is about. That is about that. This is about that which is about that. That is about that. This is about that. That is about that which is about that which is about that which is about which is that is about which is about that. That is about that which is about that which is about that which is about that which is about that which is about that. This is about that that is about that is that is that is that is that is that is that is that is that that is that is that is that is that is that is that is that is that is that is that is that is that is that is about that is what it is about. This is about this. This is about that. That is about what that is about that is about what that is about that is about that which is about that which is about that which is about that which is about that which is about that which is about that. This is about that. That is about that which is about that. This is about that which is about that which is about what is about what is about what is about what is about which is about what is about what is about. This is about what is about what is about what is about what is about what is about what is about what is about what is about what is about what is about what is about this and that. That is about the thing that is about the thing that is about that which is about that which is about that which is about that which is about that which is about that which is about that which is about that which is about that which is about that which is about that which is about that. This is about that. This is about that. That is about that which is about that which is about that. That is about that which is about that which is about that which is about what it is about. This is about that which is about that which is about

Naturalization

that which is about that which is about that which
is about that which is about that which is about
that which is about that which is about that which
is about that which is about that which is about
what is about what that is about what is about
what is about what is about that. That is about
this. This is about what is about what is about
what is about what is about what is about what is
about what is about what is about what is about
what is about what is about this that is about that.
This is about that. This is not about non-sense
logic. This is about the reality of the world. This
is about what is necessary. This is about what
logic is. This is about the hypo-real. This is
about a non-reality. This is about the reality of
the world. This is about the legitimacies of
justice. This is not about the becoming of man.
This is how the man cooperates with himself.
This is how the man operates and becomes. This
is about how the man is crucial. This is about the
self and how it will go far. This is about the
becoming of mankind and man. This is about
how man will go far. This is about the science of
man, and becoming. This is about what the man
does, that becomes. This is about science and
progress. This is how man relates. He relates to
his senses. This is about the truth of the world
and how it doesn't go astray. This is about what
the becoming of man is about and how becoming
is about God. This is about how the mankind man
goes astray. Jesus Christ is the only answer. This
is about Jesus Christ and how He goes astray. It
is about the Son of Man. This is about the only
answer to Heaven. This is through the salvation
of the cross. This is how we cope with life, and
how we keep it simple. This is through the

Naturalization

becoming of man. It is the way in which we cope with life that keeps it simple. This is through the foundation of mankind. This is about the crucial reality. It is about the becoming of man that is critical. This is about how man operates and relates. This is about the becoming of man that is critical. This is about Jesus Christ. This is about how Jesus Christ relates to Jesus Christ. This is about how Jesus Christ's secret identity was about how the world was His. This is about God. This is about how God was man and was Jesus Christ and this was His secret identity. This is about how God was in secret His whole life. This is about how God's secret identity is about the way He became through man. This is about how secret identity of God changed the world. This is about how He changed the world through Revelations. This is about the Beast, false idol worship, and world domination.

This is about how the Beast has false idol worship and a following. This is about how the world deserves much more than this. This is about the idol worship of the modern day and the way it is horrid. This is about the secret identity of Christ.

This is about the death of Christ and how He rose again.
This is about the world and how it is in critical condition. This is about the world and how it is in critical condition. This is from false prophesies and idol and Beast worship. This is about the world's condition.

This is why there is a secret identity of Christ. This is hidden from liars and lunatics. This is about the secret identity of Christ and how it is solved. This is the hidden meaning of the six,

six, six, in Revelations. This is the secret identity
of Christ and how He is this. This means that
these people are false Christ's and they create
themselves by finding the secret identity of Christ
through idol and Beast worship. This is how they
find the true meaning of Christ. This is through
the meaning of the mark of the Beast. A hidden
mark on one's hand doesn't mean that they
become this way. They *become* this way because
of the way they worship Jesus. Jesus in the form
of idols and Beasts isn't the best world. They
become this because of the fact that they *become*
the false teachers of the world. They learn the
Bible wrong to worship Beasts. This is about the
false world of false teachers. This is about how
they learn wrong and teach wrong. This is about
how they learn about false teachers and how they
become false teachers. They are the writers of all
the learning. This is why worldwide teachers are
fantastic. The teacher is not false if they are about
learning. The false teachers are about worldwide
domination through Bibles. This is about how
teachers and false people do not commit suicide.
They just go away. They go far, far, away.
This is the place they go; they go to judge. This is
how they judge.
They are judged by all their mistakes and how
they made them. This is about the Antichrist and
how he thinks he is a judge. This is about his
mistakes and how he made them. He was raised
by the world. This is how he was taught. He was
taught to do wrong by his teachers. And, what he
learned is what they taught him. This is why he is
innocent. This is about his teachers and about his
learning. He grew up in hell, and he was sent to
earth to bring people down. This is about how he

Naturalization

committed adultery in his mind. This is how he was sent to earth to bring people to hell. This is about how he can bring people to hell. This is about how he becomes. He becomes with false prophets and Antichrists of the world. This is about his story, and how it is also the bible. This is about how he uses hate to conquer enemies. And, this is how he uses lust to bring people to him. This is about how he uses devices of the system to conquer. This is about the wisdoms of the world and how it is his. This is about lies and deceit. This is about how he uses his power to corrupt and condemn. This is about how the Bible will win. This is about the tale of the Antichrist and how he grew up. Now, he is everywhere, deceiving and lying to people, of all race and color. He is about the philosophy of becoming, just because he is in it. This is about the world that is his color. This is about how he becomes with people, who do not become with him. This is about how the color of the world is good. This is about the world color, and how he condemns. This is about the world of the Antichrist, and how we can be welcomed to it if we want to. This is about the color of the world, and how we succumb to it.

An Allegory to Knowledge

"Wisdom is needed here. Let the one with understanding solve the meaning of the number of the beast, for it is the number of man. His number is 666." Revelations 13:18.

This is about the Presidential race, and how Barack Obama isn't the Antichrist. He

Naturalization

cannot succumb to this conviction. This is about
his values as a Christian and his morals. If he is
black, he can be a Christian. This is because he
can create supporters. This is people that are like
him, in Christian faith. This is like him, in
Christian belief. This is just because I am a white
republican voting black. I was a Christian
because of a black man and this is why I support
him. This is because he is about the world
campaign. He could be about world domination.
Yet, if he is a Christian, I suppose that he couldn't
start a war. This is with the world. Christians are
separate from the Antichrist, and that is the
meaning of the six-six-six. This is about Barack
Obama and how he is a President. This is about
his morals and convictions. He is a black senator
and he is a liberal. This is what a Presidential
race should be like. This is about a black
president. This is about a black liberal Senator
running for the Presidential race. This is about a
white, inferior, male, running for king. This is
about moral convictions, and how they break our
heart. This is about the king and how he long
lives. This is about myself as a king, my whole
life. This is a good story. This is about success,
and the insides of the white male suburbia. This
is what the male suburban life is about and what it
is about is male suburbia. That is all it is. This is
what the Presidential race is. This is just a
Presidential race.

        This is about the Presidency, and how this
is either black or white. This is about the modern
day. It is about this, because it has been on my
mind. This is about power, and corruption, and
how power corrupts. Power corrupts, and
absolute power corrupts absolutely. This is about

Naturalization

how corruption is in the world, and in the reality of the world. Just because we are powerful, doesn't mean we are more powerful than the common people. This is about a judgment that is not right or just. This is about a judgment that is right and just. This is about a reality that is real. This is about a reality that is real, that is about reality. This is about reality, which is real, and is about just and true reality. This is about just and true realities, which are about just and true realities. This is about the way of the world and how it is corrupt. This is because of the judgmental Senate. This is how the Senate views bills. They are for themselves only. They get elected this way. This is about how the Senate has corruption, and how we deal with it. This is about power, and corruption, and how we deal with it. This is through the common people that we step on. This is about good and the mark of the Beast.

There is no good and the mark of the Beast. This is about the mark of the Beast and how we are forgiven. This is just if Jesus Christ forgives you. This is about the mark of the Beast, and how we are forgotten. This is about the true meaning of Christ, the secret identity of Christ, and how He is God. This is about the true meaning of Christ, and His followers. This is about love, and what makes the world become. This is about Obama, and how he becomes President. This is about people, and how they make up choice. This is about the Presidential election, and how he gets a popular vote that is not from the Beast. This is about a President and how he can become popular. This is about how he can become a good thing. This is about the

Naturalization

Beast, and how he loses. This is about the mark
of the Beast, and all it is, is false salvation. This
is about how all the world is about Jesus. But, it
is better in the end to have Jesus Christ. This is
how the popular vote can win. This is for electing
Senator Obama president or John McCain. This
is of Christ. This is how a powerful leader can be,
and become a win. This is about if the man is a
Christian. And, this is not with false salvation.
This is with pure salvation. This is only for the
good side. This is called faith in Jesus Christ.
This is what winning or losing is about that is
about the world. This is about the world
government. This is about how the government
can have a world of their own. This is before the
judgment day. This is how Christ relates to
politics. This is how people relate to politics.
This is about judgment day. This is about all
being for Jesus Christ. This is about Jesus Christ
and politics. This is about a judgment day.

This is about Jesus Christ. This is all
about Jesus Christ. This is why Jesus Christ
comes. This is to save us all from sinners. This is
what we are.

This is until Jesus comes to save us. He
will come like a thief in the night.

This is about Jesus' final judgment. He
will judge all. This is all about serious problems
and politics.

This is about the mystery of the 666.

Sinners.

Sinners.

Sinners.

Beast.

Beast.

Beast.

Naturalization

World domination.
World domination.
World domination.

This is about an expression known as writing.
This is about the world that is about the world that
is about the world.
This is about the real world. This is about the real
world and how we maintain the real world. This
is about the real world, and how it becomes. This
is about real world becoming, and real world
knowledge. This is about fantasy and how it can
come true.
This is about Jesus, and the number of sinners.
This is up to Him. This is because He is God.
This is what God is about that is about what God
is about. This is about God. This is what God is
about that is about what God is about. This is
about God. God won't go away. This is why you
are sure with God. This is why God is sure with
you. This is about how God is about how God is
about how God is about how God is about how
God is about how God is about how God is about
how God is about how God is about how God is
about how God is about God is about how God is
about how God is about God is about how God is
about how God is about how God is. He is about
Himself. This is according to God. God is about
all things that God is about. This is about what all
things that God is about. This is about God and
what He is about. This is about Him. This is
what God is about that is about Him. This is
about God and what He is about is about Him.
This is about God and what He is about is about
good. This is about God and what He is about is
about Him. This is what God is about. This is

Naturalization

about God. This is about God that is about God
that is about God that is about God that is about
God. This is about how God is about God. This
is about God that is about God that is about God
that is about God. This is about God that is about
God. This is about God that is about God. This is
about God that is about God that is about God that
is about God. God is about God that is about
God. God is about God.

He is about Himself, which is about
Himself. God is a man and so are we. This is
about how God is a man and how He becomes
with all. This is about God that is about God that
is about God. God is about God. He is about all
the things that He is about. He is about God that
is about God. He is about God. This is about
what God is about that is about God. This is
about God that is about God that is about God.
This is about God that is about God. This is what
God is about that is about God. This is about
God. This is about God and about how God is
about the way He is about. This is about God.
This is about what God is about that is about God.
He is about the way that the world is about itself.
This is about the world and what it does. A
president is a president as a president. This means
that he is about himself, which is about himself,
which is about himself, which is about him. He is
officially a president. It doesn't matter what you
think of him. He is just a president. He becomes
with the people. This is how he becomes. He
becomes as a president. He becomes in this
writing form this way. This is only.

This is only what he can become. This is
a president. This is how a president explains
himself. This is through being a president. This

Naturalization

is how deformation explains itself. This is just because a president explains himself this way. Deformation is the absolute way. This is of thinking. This is in terms of thinking of it all. This is about all and if they are thinking of all then they are about all. This is about all. This is about all of deformation that is about all of deformation. This is about what deformation is that is about deformation. This is all for deformation. This is about deformation and what it stands for. This is about the way of the world. This is about what deformation is about that is about what deformation is about. This is about the world that is not about bad things. This is about God. This is about what God is about that is about what God is about. This is about God. This is what God is about that is about what God is about. This is about the God of the world that is about the God of the world. This is about the God of the world. This is what the God of the world is about. This is about the God of the world that is about Jesus Christ. This is about Jesus Christ and how He works. This is about God and how He becomes. This is about becoming that is about the way we become. This is about becoming. This is about what becoming is about and what it becomes which is about all. This is about all of becoming. This is what becoming is like and what becoming is like is about what becoming is. This is about becoming and what it is about is about becoming. This is about becoming. This is what becoming is about that is about becoming. This is about becoming that is about becoming that is about Brett Scott that is about becoming. This is about the way becoming is about the way becoming is about the way

Naturalization

becoming is about the way becoming is about the
way becoming is about becoming that is about
becoming. This is about becoming is about that
becoming is about becoming is about becoming
that is about becoming. This is about becoming
that is about man that is about mankind. This is
about what mankind is about and what it is about
is about becoming. This is about becoming that is
about becoming that is about becoming.

The Mystery of the Biblical Revelations and the
Beast

This is about what becoming is about that
is about becoming. This is about what becoming
is about that is about becoming. This is about the
power of the world that already knows what the
triple-six is. This is about the world that is about
the world that is about the world that is about the
world that is about the world that is about the
world that is about the world that is about the
world that is about the world that is about the
world that is about the world that is about the
world that is about the world. This is about the
world. This is about becoming that is about
becoming. This is how the triple-six works. This
is through the thought of becoming that I know
this. This is about mind control and Beast
worship. The worship of the Beast creates mind
control. This is of real matter and energy. This is
through the brain that creates it. These people are
mind control freaks. This is real energy from all
that creates it. This is what the real mark of the
Beast is. This is a symbol of mankind. This is
how he was enslaved by the Beast that never goes
away. This is what Revelations is for and about.

Naturalization

This is about the Beast worship that controls the process of thinking that controls you. It is a choice you have that controls you. This is whether it is from the mind or the body. But, both the hand and the forehead is marked. This is about what the Beast offers that is about the Beast. This is mind control.

If you have lost your power of Jesus and Satan then you are hopeless and you turn to the Beast. This is about what the Beast offers that is about how the Beast becomes with man. This is how whether or not you are a Christian. This is about what the Beast is about that is about the Beast. This is about what the Beast is and what the Beast is about that is about the Beast. He is about the world. This is what the Beast's explanation is in deformation. The Beast is the Beast as the Beast. This is about how he controls people and manipulates them. This is by having followers in which are almost perfect their whole lives. They look like this. This is about hate that is violence that is hate. This is in the language of deformation. This is about how hate creates people to want to worship the Beast. This is about the mark of the Beast that is about the mark of the Beast, that is about the mark. This is of the triple-six. They call it the mark of the Beast. This is about how we understand it. This means that there are sinners in the world.
The triple-six controls this. This is a meaning beyond reason. It is just a number.
The triple six is a number that is a number that is a number. This controls minds. This is about the number that controls minds. This is how it controls minds.

Naturalization

It is about other people's money that is about other people's money.

This is about relating sexually on the Beast. This is through corruption.

Some people "relate sexually" on the Beast by worshipping him.

Others, "relate sexually" on the Beast by falsely saying he is God. This is about the world and what it represents. You are either a follower of the Beast or you are not.

This is about the triple-six.

This is a mark symbolizing the relationship between man and God. This means that if you do not worship God you worship the Beast. This means that the triple-six is about creating things that are not yours. This is by creating violence when it is not yours. This is about creating hate when it is not yours. This is about the meaning of the Beast, that is the meaning of the triple-six. This is about what the six means.

It is a number in ancient numerology that represented Gods. These are what they mean. They mean that they are discovered this way. False God's are discovered through the six. This is about the discovery of God. This is through the mark of the Beast. This is about how it is all lies, corruption, or hate. This is about what the six represents and what it represents is all hate, violence, and greed in the world. This is about values and naturalization. This is about how the triple-six agreed on doing this. This is about how the triple-six becomes a number. This is through the word of God being manipulated by people. This is about writers and readers relating. If there were a triple-six on this book, they would relate

Naturalization

falsely. They would probably go to hell. This is according to the book of Revelations. The reality of the mark of the Beast, is about hell. Hell created it for people who wanted to follow him. This is about God, and how it is a number against Him.

This is about persecution and the church. This is about the meaning of the triple-six. This is about how people can relate to the way it is put on their head. It is put on their head by the Antichrist. He is the symbol of hate and violence. This is about the triple-six and how it has mystery and allure. This is about how it is about how it is about. The triple-six is about the triple-six which is about the triple-six which is about the triple-six, which is about the triple-six, which is about the triple-six, which is about the triple-six, which is about the meaning. This is about the meaning of the triple six. It is a device of the system. It is about the way we become.

We become through violence, hate, and anarchy. This is about how the triple-six relates to the world. This is about how violence, hate, and anarchy aren't natural. They are designed by the reality of the world. Some call it the system, some call it natural design, yet others call it a necessity. This is about how we think.

If the Bible thinks we can have a triple-six, then we can have a triple-six. A hell would be created by it, if worshipped this way. This is about intelligent design and if there is any. This is about the design of the world and if there is one. This is about the design of the system. This is about the design of the world. This is about the design of the matrix. This is how love interacts with violence. This is about the mystery of the

triple-six. This is that it is a mystery behind it.
This is about being able to control people and
create a world. It is about the mystery that is
about the mystery. This is about the mystery that
is about the mystery. This is about the mystery
that is about the mystery. This is about the triple-
six. If it were solved, there would be no more
violence. This is about the changing of reality
and the matrix code through the reality of the
Beast. The mystery behind the triple-six is about
the mystery behind the triple-six. The only art-
form smart enough to explain it is deformation.
This is about the way it is explained. This is
about its explaining. This is about the way you
explain it that makes a difference. This is about
the creating of the world to be just yours. This is
about mercy and how it has none. This is about
an explanation and how it has one. This is about
the way of the world and how it has none. This is
about the world and how it has will and
representation. There is a world with
representation that is about the real world. This is
about the world that is about the real world. This
is about reality and representation, and how there
is none. But, there also is a reality with
representation of the triple-six. This is about the
mind control and how it has policies. This is
about how you worship the Beast. This is about
how worshipping the Beast is about nobody. This
is why there is no reality to it. This is about how
reality interacts with the triple-six. This is about
the reality of the triple-six and how there is a
mark on some people's heads. This is about
predestination and how there is none with it. This
is about fake mind control and how there is none.
There is always a mark of the Beast controlling

Naturalization

someone. This is how this mystery is never solved. This is only solved through math. The number that represents the triple-six are about zero and one. There is a number chart. Zero and one make up the number of the Beast. This is because the zero is the Beast's number, because it came from hell. And, the one is the number of the people who are involved. There is one God, that is them, and a false God known as a symbol that is the Beast. The number of the Beast is solved by this equation.

$0 + 0 = 1$

The Beasts that are two are numbers 0 and 0. The number of the mass that follows is 1. The numeral zero is about the way the Beast relates to people. He relates through the world's becoming. This is because he was fortunate enough to be smart. The numbers represent science and the big bang. This is about the formula for the big bang. It is about the the world's becoming that gets us. Math-words is a solution to this. If the Beast were zero, then he would be a loser. If the Beast were zero, then he would be a winner. The one symbolizes all the people who go to hell. This is because of the way the zero works. It is about being placed on the forehead, as a six. The other zero is about being placed on the forehead as another six. The one is about the following of the zeros. This is about the way the zero is the zero that is the zero. This is about the Beast. This is about how the other zero is the zero that is the Beast. This is the other Beast. The Beasts are symbols of 1. Idol worshiping, and 2. Beast worshiping. This is about the zero and how he is the Beast. He came from a zero. It is just like getting nothing for a credit on a test. This is about world without representation. This is about an empty world. This is where he came from. This is about false Beast worship and how it came from nothing. This is about hell and where they go. This is about idol worship and how there is none. This is about the mark of the triple-six and how it sends people to hell. This is about deformation and how it explains it. This is a meaning that means that the Beast is nothing and the world is something. There is a meaning saying that if you worship the Beast, you go to Heaven. This is all about false. The real meaning of the Beast is meaningless. This is because he is a false God. This means that there is no meaning behind the triple-six. This is about the Beast and how he has no meaning. This is about zeroes and how they mean something. This is about the continuance of thought and how it means this. But, thought is just one creation of the Beast. This creation of the Beast, is about the Beast, and how he means something. This is to a lot of people. Yet, their number is one, and the Beast is zero. The Beast has meaning beyond knowledge. Worship is defined as worthiness, respect, or reverence paid to a divine being. This is what the number of the triple-six is not. This is about the manipulation of the number system. This is because zero is not a number. This is because the real number of zero is nothing. This is what the triple-six is. It is symbolic for no meaning. Good is good as good. This is about how good is good. The number six, six, six, is about the great and mighty one. He is almost a divine being. He is about hate, violence, and greed. The number of the triple-six, is about the hatred of mankind.

Naturalization

Deformation explains that the mark of the triple-six is inherited. It is given to you somehow. This is just by a choice. And, like a toilet paper roll, it exists. This is about the divine wisdom of God and how it is manipulated. This is about how sinners go to hell, and it is a way of doing this. This is about the world that is about the world that is about the world that is about the world that is about the world that is about the world that is about the world that is about the world that is about the world that is about the world that is about the world that is about the world. This is about the way the world works. The divine number of six-six-six comes from Heaven. It is about the fallen people and how they go to hell. This is about the understanding of people. This is just what hell understands. This is about the relationship of the Beast to his people. They are about relating with the triple-six. This is how they have love and compassion towards each other. But, they are hypocrites. Love is war to them and violence is compassion. This is only to people outside their group. They have divine wisdom that creates the six-six-six because they are the only ones who relate to this. This is the mark of the underworld. Like, there are six chambers to a pistol, there are six numbers to the six-six-six. But, this is what the world wants to know. This is about wisdom. Even, with the numbers the people are still against wisdom. So, this does not help. The number chart would say that there is a problem of reason with it. This is why it is not a number. It is a language. The marking represents how the language works. It works for losing. This is why they all go to hell. This is about how the world relates to the world.

Naturalization

This is about losing and how they lose. The hell
is open to anyone who has the six-six-six. This is
about understanding with divine wisdom. This is
all the six-six-six is. It is an observant person. It
is a scientist. This is why they have divine
wisdom, which is the only way they survive. So,
they are destined to go to hell. This is about
people who go to hell. This is about the world
and how it represents itself. This is about how the
people of the world are all destined. This is about
the world that is about the world that is about the
world. This is about the world and how it is about
the world. This is about how the world is about
what the world is about. This is symbolic of the
Beast. He moves through the number triple-six.
If he wanted someone dead, they would become
him. This is about how there is six of them and
the last one is like a God. This is about thought
that is about what thought is about that is about
what thought is about. This is about God. This is
what God is about. He is about the things that
man are about. This is about man and how God is
about him. This is about what man is about and
what man is about is about man. This is about
how The Antichrist just manipulates the system.
He is the one to worry about. He does not accept
anyone. He is the only one who would know
what the triple-six is. It is about the people in the
world and how they are about him. This is about
how he knows this so it is different from him.
This is about how the world is about good and
how the good of the world is about is about hate.
This is with his knowledge. This is about how the
sixes represent people who want to worship the
Beast. This is about how the world is about this
in the future. This is about how the world will

Naturalization

one day defeat the Beast. This is about the world and how they prepare for him. This is about the world and how people are about it. This is what the world is. It is a hunting ground for the Beast. This is about divine wisdom and how they have it. This is about being smarter than most things. This is about the world and how it has fallen. This is about how divine wisdom won't understand the Beast. This is because if it understood it, it would lose. This is about knowledge and how it has an answer. Anyone powerful enough to power the Beast must be him. They are just like him. This is just because. They have no reason. They just become with the triple-six. This is eternal. And, they also have power over evil. They can do miracles, and move the divine church. This is about how they win. It is hell. This is about the world and what the world does that represents the world. It doesn't represent itself, or we would all die. So far, it has been evil. But, there is a just improvement here again. It is about Jesus. Jesus Christ rules the world. Jesus has just anger at these people. But, He will return to save us. This is even if the Beast is worshipped as a false God. This is how the Beast will return. This is just if people worship the Beast. This is what God is and what God is, is God. This is just according to the Beast worshippers.

The theories of naturalization are right. This is about deformation. This is how a theory explains how we can become with God. This is through the following. This is not about the beast, but about God. This means that there are no Beast worshippers.

There is only the world. This is what the world represents. This is about the world of Beast worshippers. But, to really be smart you must worship God.

Antichrist

This is about the king of lies and deceit. This is about unusual occurrences. This is about how one man rules the world. This is about how one man can corrupt or destroy anything. Yet, he is an angel. This is the angel of the Lord. He fell from the Garden of Eden. He is really a serpent. Some say that he is an angel. But, this is wrong. He is a Serpent. This is what he was in the past. In the future, he is an angel. The church knows how he goes out and deceives the world. This is why they are so scared of him. This is because he is an angel. He can take on different forms and become other people. This is just with one power. But, he is fictional. He must be created, to exist. Yet, he can create himself in realities. He has power over reality just by becoming it. And, he is the real McCoy. He is about abuse and violence. Heaven created him, just because he fell one day. He was about sex, and violence. Yet, all he was about was being better than God. This is why I have heard he has fell. Yet, other people have other stories. They have heard about him differently in church. This is according to subjectivism. This is just about lies. This is just about deceit. This is about the collective intelligence and how people try to be this. But, there is one Antichrist, and he does rule the world. This is just by deceiving everyone. The world was created by Antichrist, according to him. But,

Naturalization

there is a hell.   This is how lies work. They work
through things that work through things. It is
about lies and truth that is about lies and truth.
This is about truth, that is about truth. This is
about lies that are about lies. This is about truth
that is about truth. This is about lies that are
about lies. This is about the world that is about
the world. This is about truth that is about truth.
This is about lies and truth and about lies and
truth. This is about truth and lies and about lies
and truth. This is what truth and lies are about.
This is about truth and lies. This is about truth
and lies that are about truth and lies. This is about
truth and lies that are about truth and lies. This is
about truth and lies that are about truth and lies.
This is about lies. This is about how truth is truth
and lies are lies. This is about truth and lies. This
is about truth and lies. This is about truth and
lies. This is about lies and truth. This is about
truth and lies that are about lies and truth. This is
about truth and lies. This is about truth and lies.
This is about lies and truth that are about lies and
truth. This is about lies and truth that are about
truth and lies. This is about lies.

The Antichrist is about lies. He is about
truth that is about truth. This is about lies that are
about lies. This is about the truth that is about
lies. Lies are about the truth that is about lies.
Lies turn the truth into lies. This is about lies and
truth and what they get. This is about truth and
lies. This is about truth and lies. This is about
lies and truth. This is about lies and truth. This is
about lies and truth. This is about truth and lies.
This is about lies and truth. This is about lies and
truth. This is about lies and truth. This is about
truth and lies. The truth, according to lies, is the

Naturalization

truth. Lies, according to truth, are lies. Lies are about truth, that are about lies. Truth, is about, truth, that is about lies. The truth is different than lies, that is about the truth. Truth, is about lies, which are about truth. Lies, are about lies, which are about truth. Lies, are about truth, which are about lies. The truth, is about lies, that is about the truth. Truth and lies are separate though.

This is about truth. This is about lies. This is just one part of Antichrist's infinity. He is like an angel with any power. This is how evil works. This is about how he deceives and lies to us. This is about how the cops are corrupt. This is about how the skull and bones are a cult. He deceives and lies, to us collectively. He is the only thing that can do this. That is about the only thing that can do this. That is about the Antichrist's lies. He is merely an angel with powers of God. This is why he thinks he can win. This is all about lies. So, he doesn't win.

He becomes chained to a thing for the rest of his life. And, he is an angel, so it is like eternal life. This is about how he lies. He lies to get advantage of others. This is about a lie, so it is about a lie. This is about a truth, so it is about a truth. This is what others, see, so what others see. This is about what they see. We see in terms of shapes and colors. This is how he manipulates. He causes hate and violence. He causes things in which send people to hell. This is about manipulation. He is about the world, which is about the world. He is about an angel, which is about him. He is about an angel. He is about himself. This is what he does. He goes out to deceive the nations. He goes out, to become with the world. He is about lies, and truth, and how

Naturalization

lies and truth overcome. This is about Antichrist, and how he overcomes. This is through the world's becoming in addition. This is about the satanic church becoming. This is about evil overpowering what is good. This is not about the satanic church becoming. This is about evil prevailing. The Antichrist does this. This is about the Antichrist becoming what he can become. This is because he is about lies that are about lies that are about lies that are about lies that are about lies that are about lies that are about lies that are about lies that are about lies that are about the hypocrisy of lying. This is about the all of all. This is the world that is different than me. This is about the world's becoming. The Antichrist doesn't become with the world. It is better than me. It is about the world that is about becoming, that is about becoming, that is about becoming, that is about becoming. This is about becoming. This is about overcoming the world that is about the Antichrist. The Antichrist overcomes the world. This is the world that the Antichrist created. This is his lie. This means that he is insane. He is the hypocrite of the world. This means that Antichrist has no truth. He is about becoming with people who become with him. This is about the truth and how it is manipulated. This is about how people like to become with other people. This is if they changed their mind. The Antichrist can change their mind. This is about the world and how we do it. This is about the world and how we do it. This is about the world and how we operate in this. This is about the world and how the world operates. This is about the world that is about the world that is about the world that is about the world, that is

Naturalization

about the world. This is about the world that is
about what the world does that is about what the
world does. This is about the world. This is what
the world represents. This is Jesus Christ. This
is what Jesus Christ does. This is very important.
This is becomes with others. This is about the
Jesus Christ of the world. This is about how we
relate. This is about the way the world works.
This is about the world that is about the world.
This is about the world. This is how Jesus Christ
works. This is about how the world works and
how the world works is how the world works.
This is about how the world works. This is about
how the world works. This is about how the
world works. This is about how the world works.
This is about it. This is about it. This is how the
world works. This is about how the world works.
This is about how the world works. This is about
how it works and how the world works. This is
about Jesus Christ. This is what the Antichrist
hates. He hates Jesus Christ. This is because of
his demeanor. This is how the Antichrist works.
He thinks that all people are different when we
are all just the same. This is about how the
Antichrist works. This is all about how the
Antichrist works. This is about him that is about
him. This is about the Antichrist that is about the
Antichrist. This is about how he works. He
works through the world. This is how he works.
He works for people and for evil. This is because
the Antichrist is among the most evil of all. He
works for Satan. This is because he is Satan. I
am a writer who likes this. This is just because
my intellect is curious about Satan. This is about
how Satan works and operates. He is the one who
makes the intellect curious about him. This is just

Naturalization

by existing. This is how Satan exists. He exists in a world of illusion. This is because he can change forms. This is how he works and how he works is through himself. This is about the art-form known as deformation. This can explain him.

He is an angel that is about an angel that is about an angel. He is about evil that is about evil. This is about evil that is about evil.

Evil.

Evil.

This is about how Satan works and operates. This is about him. This is about him only. This is about a loaner that is about God. He is about God because he fell. God is his worst enemy. This is about deformation that is about deformation. This is about how Satan is Satan. Satan is a genius. He is actually anything he wants. This is in the context of evil. This is about Satan and how he wants everything. This is about everything and how he wants it. He wants everything.

His secret identity is the triple-six. He is the one who created it. This is because he is miserable. This is how he is miserable and how he is miserable is how he is good. The world hates Satan. And, this is how Satan survives. Satan survives through the intellect he has. This is about trying hard. This is how Satan tries. This is only for evil. This is how Satan does it. He does it only for evil. This is about how Satan is evil and survives. He survives on bread alone. He probably doesn't have to eat anything or sleep. This is because he is a creation of the world that he is in. This is about the world and he is in. This is about evil. This is about pure evil. This is

about pure evil. This is how pure evil works.
This is about pure evil. This is how evil works
and operates. He is about pure evil because he is
about pure evil. He is about pure evil that is about
pure evil. This is his identity. This is his secret
identity.

God's secret identity is Jesus Christ. This
is how evil works.
It works opposite of what Jesus Christ wants.
This is because Jesus Christ is pure good. This is
how Jesus Christ makes it. He makes it a good
world unless Satan wins. Satan sometimes wins.
He is about pure evil. This sometimes wins in
your life.
Deformation.
Deformation.

Deformation is about the art-form known
as deformation. This can explain anything and
anyone. This is about the world known as God's.
This is about good that is about good. This is
what it is. It is an art-form. It is about the world
and about the world. This is how it
communicates. It communicates through Satan
even. This is because it is about words. This is
about communication that is about
communication. This is about deformation that is
about deformation. This is about the world that is
about the world. This is about deformation that is
about deformation. This is about evil that is about
evil. Deformation only thinks of evil because of
Satan. This is about how deformation
understands itself. Deformation understands itself
through everything. This is because it is about
itself and is about itself. This is about how
deformation is about how deformation is about
God. This is about God. Deformation

Naturalization

communicates through the unknown. This is called reality based knowledge. This is about the world that is about the world. This is about God that is about God. This is how the world is about the world. This is about how the world is about the world. This is about how the world is about how the world is about how the world is about how the world is about how the world is about how the world is about itself. This is about the world. This is about how the world exists that is about the deformation artform. This explains everything through the world. This is about how knowledge can explain everything you want. This is about how knowledge shapes your life. This is through the world. It is through the world that you want everything and get everything. This is through the world that is about the world. This is about the world that is about the world that is about the world. This is about the world that is about the world. This is about the world that is about the world. This is about the world that is about the world. This is about the Antichrist and his number. His number is six, six, six. He is the one who rules the world.

This is about knowledge and how it becomes. This is about the beast and his number. This is about the beast and his number. His number is similar to God's number. But, God does not have a number. This is about Christianity and how it rules the world. Christianity did not create the bibles, Jesus Christ did. This is about how the world does it for Jesus. This is about an adventure known as man.

This is until the Antichrist came. He is the one who rules the world.

Naturalization

He is the real man who no one will know who he
is. This is about him and about his knowledge.
He is a king in business and a king in industry.
He thinks he owns the world. Yet, he is very
meek and modest. I am not the Antichrist writing
this to you. I am a saved Christian. He works
through wisdom and knowledge.

Yet, he does not have the belief system of
Jesus. He thinks he can order Jesus to do
anything to you. This is about how he becomes
and how he becomes is with people.

This is about how he tries to become with
people. He tries to become with them and be like
them. This is how he tricks people. He is the one
writing this book to you.

This is because he is Jesus Christ. He is
the one who makes Jesus Christ look stupid to
everyone. He is also a lawyer, a banker, a
mathematician, a skateboarder, a cop, a business
tycoon. These are the kind of people he controls.
Yet, his number is 666 so he is not me.

He just thinks he rules the world. He
thinks he becomes with people. This is about his
theology.

He thinks he can be like Jesus Christ. Yet,
he is just the opposite. This is how he dominates
the world. He has a reality that is like Jesus
Christ's. He thinks he is like Jesus. But, he really
is not. He just becomes with Christians. This is
because he is an evil spirit. He can become with
anyone and everyone. He thinks he is God,
because of the way he acts. He is really just a
humble servant known as Jesus Christ. This is
because he crucified Christ. He is about making
people like himself. This is about the Antichrist
and his return. Yet, he cannot come to people like

Naturalization

a thief in the night. He becomes through Christians. This is because he knows Christianity well.

He is about the things that are about the world. He is about hatred and violence. This is toward all women. It is also toward all Gods. He is the one who created the false Gods and worshipped them. This is about the beast.

This is how he was not created. He is in a book called Revelations.
So, is the Antichrist. He is in there as Satan. He goes out and deceives the nations, as well. This is about how the Antichrist is the pure Satan. He is like a ghost in the night. The beast's number is six, six, six. This is just how Satan becomes.

He is like a Christian. He is like a Christian, as well as a lawyer, and banker, and mathematician, a skateboarder, a cop, a business tycoon because he isn't like them. He is a heretic and lunatic. He is all these things and more.

He is about the math that is about the world. He is about adding up the triple-six. He is about the creator of the triple-six and how it happened. It is weird. This Antichrist becomes with people everywhere. He is about the world that is about the world that is about the world. He is about the world that is his. He is about hate and violence. He is about becoming with man and woman. This is about how he does it, and how he does it, is through business deals. He is about you and about the real world. This does not respect him.

He is a copy-cat, con artist. This is about how he takes on many different forms. This is about reality and how he is about it, is about him. This is about how he takes on many different

Naturalization

forms. This is about the world in which he lives in. This is about pure world. This is about how he controls the world and manipulates it. This is about how he becomes with you, to exist. He exists, and how he exists is through the world's becoming. This is about the world, and how he becomes. This is about cons.

This is how cons work. He is about the world and who made this happen. God created the world. This is about how God created the world, and how it became. He fell as a Serpent as the world was created. This is our world. This is not the Garden of Eden. This is where he came from. He is an angel who says he is a serpent. This is about the world and how it fell. This is because of him.

He fell from Heaven one day. This is as the serpent fell. This is what he chose to be. This was an angel. This is how he deceives the serpent. This is by trying to become with him too. This is about lies and truth. This is about the world and how it has truth. He is blind to the truth, so he is not a Christian. He is a real guy who rules the world.

He is behind the mystery of the six, six, six. He is behind the becoming of the world. He is behind every cop, banker, and lawyer. He is about becoming with people, and how they become. He is about being as smart as God, and as wise as Jesus.

This is about his hate scheme. This works through hate, through the world. He has pure hate. This is opposite of love. This is how he becomes.

Naturalization

This is all that he can do. It is work through hate toward Jesus. Then, he tries to become through God.

This is because he is the God who hates himself. This is how he does it. He does is through hate. It is how he becomes. If there weren't becoming. Then, he would not be able to exist. This is why becoming is so important.

This is how he exists. This is through the world of thought and knowledge. This is how the triple-six works. This is like a false-God that rules to your knowledge, thought, and creativity. It is about a false God and how he worships. This is about the thought of thinking. This is about how he thinks this. He thinks as you, which is why he exists. He thinks in terms of beast worship because he is about all. This is about all hate, violence, and anger. It is how he becomes which is scary. He becomes through the beast. This is all he has.

This is why they are created. This is because of the mark of the reality of the becoming of the evil that is the Antichrist. He becomes through you. He is about becoming man and woman. This is why some think he is gay. This is because he is just like you. Yet, he is the opposite. This is about the antichrist and how he tricks us. This is about beast worship. This is the opposite of God worship.

So, it really is evil. This is about the way evil works, and how it works. This is about caring and sharing. This is about love and truth. The only reason why this is evil too, is because of the Antichrist. He tries to turn love into hate, and hate into love. This is about Jesus Christ. This is how He is saved.

Naturalization

The Antichrist cannot do it to saved
people. It is how the Antichrist is God that is
about the way we love. This is in a weird way. It
is because of him. He is about love, and violence
and what they are about. He is about hate. He is
for the world that isn't for him. The mark of the
beast is opposite of the Christ. It is about food
and water. It is about the opposite of what the
Antichrist eats. He feeds off the bodies of natural
people. This is how he becomes. He eats the
flesh of man.

This is why they called him the beast. It is
about man who he is opposite of. This is how the
Serpent fell once to create all of this. It is how the
Serpent fell. He is the one who created
knowledge. He is the one who created knowledge
how to become. It is about the world that is about
the world. He is about the Antichrist. The
Antichrist is about the Antichristian man. He is
about world domination. He is just like this,
because all are like this. This is about man that is
about man that is about man in which is about
doctors, lawyers, scientists, bankers, college
professors, people who work, people who don't
work. This is about man and how the Antichrist
can become him.√

This is by being an Antichristian man. He
does rule the world. He plots to kill people. He
plots to become people. Then, afterward he tries
to send them to hell.

This is about how the serpent fell. This
was from the Garden of Eden.
This was to create knowledge. The Antichrist
becomes through knowledge. This is because he
is the triple-six on foreheads. He is the symbol of
the triple-six, as man. He is against man, which is

Naturalization

why we understand it. Yet, it is a murder symbol.
This is a symbol of justice. He is the hate man.
He is the one who understands this. This is
because he is man, and the triple-six, is his
number. This is about God and how he
understands Him. This is about how the
Antichrist rules the world. This is through the
world, that is the world, that is the world that is
the world that is for the world, which is for the
world, which is for all. This is about the
Antichrist. He will go away when Satan deceives
the earth. This is about the fall of man. This is
about the Serpent. The Serpent known as Satan,
fell from the Garden of Eden. Some describe him
as an angel. Others find meaning in deformation
as all of this, in simple terms. This is about
deformation that is about deformation that is
about deformation that is about deformation that
is about deformation that is about deformation
that is about deformation that is from this. This is
from pure understanding and wisdom. The only
reason why man fell was because of the Serpent.
He is the smartest of all animals. It is about the
world why man becomes with the Serpent. This
is because the Serpent was an animal. This was
also once described as an angel. It was probably
the Serpent's deceit that controls us. This is
through the knowledge of man. This is about how
knowledge fell. This was through Adam.

   The Serpent is the one who deceives us.
This is if the Serpent is real.
The Serpent could still walk the earth. But, it was
cursed to crawl on his belly. This was because it
made Adam eat the fruit of knowledge. This was
due to the fact that people are like people, who are
like people, who are like people, who are like

Naturalization

people, who are like people. This is about the Antichrist and the Serpent. They are very similar.

The Serpent becomes with man, just because it made him fall. This is the Serpent's world according to God. The Serpent was an angel who is Satan.

This is just because it could think that it could be smarter than God. This is about how the Serpent works. He is Satan. This is how he has to work, through lies and deceit. This is about how he cannot spell or write. How this is, is about the becoming of man. When, Adam fell, it was through becoming. He became smarter than God. How he got kicked out of the Garden of Eden was because of this. He became with the Serpent. This is how the angel known as Satan knows it. This was because the Serpent was the angel known as Satan. This is how he lies. He says he is God to people.

This is about how knowledge and him work. This is through necessity. When we fell, we started to become necessity. This is how necessity works. This is through the Serpent. He became an angel somehow. This was through transformation. This is how becoming became with Satan. He tricked Adam and Eve into biting the apple. This is how it happened. This is through sin. This is what knowledge was. It was sin. This was how He fell through the Garden of Eden. This was before time and space. This was after God created the world. This was through real justice. He was forced to walk the earth, or "crawl on his belly". But, He is pure lies, which is why we don't believe him.

This is because Satan is a snake. He is also an angel known as Satan.

Naturalization

This is why we believe in Him. This is because he fell from the Garden of Eden known as Satan. He was a Serpent. This is why some think the Bible is an allegory.

This is because of the Serpent. He is about the world that is about the world. He is about deceiving and lying. This is how the Serpent made a comeback.

He will try to deceive the world and destroy it. This is after the fall of man. This is immediately after. He is not in time and space. This is how Satan works.

Right after he fell, he will try to deceive the world. This is about how he will deceive the world. He will do this right after Adam fell. The Serpent was the most cunning of all animals. This was because he was a Serpent. This was because he fell as an Angel from heaven. This is out of time and space. This is about how Satan deceives the earth. This is about the earth and how he "walks" it. He is an angel, which means in Bible language, a snake. This is because he fell when Adam fell. He fell from the Garden of Eden, which is Paradise. This is a word for Heaven. This is how angels fall from Heaven. They fall by tempting man.

They fall from Heaven hard. And, they try to win by calling themselves angels. This is about how Satan is the king of lies, and how he is the Antichrist.

The Serpent became the Antichrist when Jesus came. This is because he tempted man to fall. This is how he tempted man. He tempted man to have God's knowledge. This is why the Serpent still does this. This is through knowledge. When Jesus came, his knowledge switched to Antichristian knowledge, because Jesus became God.

God hates Angels that lie. This is what Satan is. He is about lies and deceit. This is why he chose to be an angel. This was because he says he was God's right hand man to people. But, he really just was a Serpent. This is how He creates himself through knowledge. This is by lying and saying that he is an angel. But, a snake before time and space is an angel according to some. This is about deformation and how naturalization works. It lets you understand the knowledge that is real. The Serpent is about lying and deceiving that is about the world. He takes on many different shapes and forms. This is by crawling on his belly. This is about Satan and how he is a snake.

This is one of his identities. This is about how he cons the world. This is about lying about his true identities. This is how the world represents itself. The world comes as it should. This is about the symbolism of the end times. If you do not worship God, you worship a beast, and if you fall into temptation, you become an idol worshipper. This is about a test to make it to the Garden of Eden. This is why Satan goes out and deceives Gog and Magog. This is because the world was deceived by him when we fell. When we almost go back to the Garden of Eden with Jesus, then we must become with Satan first, which is how he fell. He deceived the world. This is about Satan and how he has superhuman strength. Only a snake would believe this kind of stuff. This is about evil and how it manifests itself. This is about belief and how it becomes. It becomes from the Garden of Eden. This is how

Naturalization

the Serpent fell. This is because he "tried to be smarter than God" in a literal way. He tried to outsmart God, and make man fall. And, he did. He did this by tempting Adam. Adam fell because of Satan's temptation. This is about the Serpent and how he deceived the world. This is about Satan and how he tempted Adam. This is because the Serpent was Satan. The Serpent was Satan. But, God was so scared that Satan was an angel that fell, that it went into the Bible as snake. This is what it could have looked like. This is to me. I am not an angel known as Satan, and I will not call myself him. This is because of my beliefs. Serpent is smart and cunning. This is how he learns. This is through tempting Adam before the fall. This is what he still does. He does this for the rest of creation. This is how man fell.

He was in Paradise. He was tricked into falling. But, the angel known as Satan, had to outsmart him. This is how Satan became smarter than God and did this. This is through tempting Adam. But, temptation is a word we know. It could have been inverted in Heaven. Words are a product of the tree of knowledge. This is how they work. They work by being smarter than God. This is just what the Serpent thought. The Serpent isn't Satan, and he also is. This is because the Serpent can change forms. This is throughout the world. This is how he walks the earth. This is how he walks the earth. This is through the world. This is what he thinks. He thinks he is smarter than God by being an angel.

This is not an allegory. This is true.

Allegorically speaking we are in a cave. And, knowledge is the light from the cave. It

Naturalization

takes form and shape behind it. This is about how it takes shape and form behind it. This is what Plato thought. We are blinded to the light unless we become free. The light is knowledge that we don't have. This is pure light. But, unless we are out of the cave, we are destined for light. If we break free, then we can become outside of the cave.

Allegories are like this. This is how Satan is. He is in the cave until he discovers himself too. This is what Satan looks like. He looks like a bright and shining star. But, deceit and lies cover him. This is about Satan and how he looks. He looks like a bright and shining star. This is because he is trapped in the cave the most. This is how the world works. This is about the cave and how it is an allegory. It is a cave because it is a cave. It is a cave, that is a cave, that is a cave, that is a cave, that is a cave that is a cave, that is a hidden place. Knowledge is about the cave. It is all about the cave. This is hidden from eyes. And, the light is the outside world. It fuels the fire inside, the cave. It is about the cave, which is about the cave. The knowledge from the outside world is entered into the cave. It is how we become. So, to take out of the cave, we would have to use knowledge from it. This is how we enter the cave. We are born and become in it. This is in the real cave. This is how knowledge escapes from the cave. This is through thought that goes out and comes back in to fuel the knowledge in the real cave. This is through entering the cave do we become born. This is about the cave and how it is a ritual for knowledge. This is how the ritualistic cave becomes. It fills itself with light, from the

Naturalization

knowledge. This is how it fills up with light. This is from outside the cave. Knowledge from outside the cave, enters in, and fills it with light. This is if you believe. This is what the fire does. It creates shapes and forms on the wall. This is what we see. This is how we have knowledge. This is how we survive. This is through picture and items on the wall to look at. But, we are like we are in a cave.

This is until Jesus sets us free. This is about the cave, in which we dwell in. This is about the cave that rituals take place in. This is what we have to do to be free. We have to continuously believe in the rituals that we do. This is what creates the knowledge around us, for all. This is how the cave is free. This is from our own behavior. This is how the cave becomes free. We go out of it; it doesn't go out of us. This is how we act that matters. This is about becoming and how we become it. It is from the light. This is what keeps on keeping us well. This is what the light does, that becomes with the light. This is sets us free.

The light that sets us free is this light. This is the light of the cave. The light of the cave is what sets us free.

It is after we become with it. This is how the Antichrist does not become with it. He is the reason we are in a cave. If we weren't in Heaven, and then in the Garden of Ede, then we would not be free. We fell from earth, to earth, and we keep on falling. "For all have sinned, and have fallen short of the glory of God" Romans 3:23. This is an interpretation meaning that there is no man who cannot fall because of Adam. But, we keep on falling, which means that we keep on sinning,

to become. This is how the snake tricked God.
This was by outsmarting him. This is why he fell.
He fell because of the Serpent's temptation and
how we believed in it. It is about the cave, and
how the cave, is about the cave, is about the cave.
We believe in knowledge, outside of the cave, but
nothing is in it, but light. This is from the fire
behind you.

This is how shapes and form is created for
you to see. Yet, the cave is living. This is how
this is. This means that the cave was created by
you. This is through thought and knowledge.
This is from the fire behind you. This is how the
fire becomes with you.

It is about the cave, and how the cave, is
about the cave, is about the cave. We believe in
knowledge, outside of the cave, but nothing is in
it, but light. This is from the fire behind you.

This is how shapes and form is created for
you to see. Yet, the cave is living. This is how
this is. This means that the cave was created by
you. This is through thought and knowledge.
This is from the fire behind you. This is how the
fire becomes with you.
This is how the fire becomes with you. This is
what the fire becomes, that becomes with the fire.
This is knowledge. This cave is living. This is
because you are inside it. You have knowledge
from the outside that goes in. This is how it takes
shape and form. This is how the ritualistic
knowledge of the cave is created. This is about
the cave that is about you. This is about how your
personality is shaped. It is how your world is
shaped. It is how good and evil do not mix. The
cave is about becoming man and woman together.

Naturalization

It is about how knowledge is the light outside the cave. And, to get it, we must follow the rules of the fire. This is about how the fire, is about the cave, which is about the cave, that is about the cave, that is about knowledge. This is just a knowledge statement in which the cave would have. It is about light and darkness. The cave would have light. The cave would also have shapes. This is the Plato version of the cave.

Chapter 6
Belief

Belief is belief as belief. Belief is about believing that is about believing. This is about believing in God. It is about God that believes in God. It is about God that believes in God. God believes in God. This is how belief works. This is through the world. This is what controls us. This is how the cave is superior. It is about knowledge that created it. This is what is about the cave, which is about the cave. This is about everyone ever being in the cave. This is about belief, which is about belief. This is what belief is about that is about belief. This is about belief. This is about belief. This is what belief is about, that is about belief. This is about belief, and why we believe in it. It is about it. It is about it, it that is. It, it is. This is about what belief that is about belief that is about belief. This is about belief, and how it works. It is about a belief in the Father, and the Holy Ghost. This is about Jesus also. This is about how belief is about the belief of the Father, Son, and Holy Ghost. This is about also about becoming. This is about becoming, that is about belief, which is about belief. This

Naturalization

belief is about the belief of the Father, the Son, and the Holy Ghost. This is about the belief in God. This is how the belief does it, which is about God. This is about the belief of God, and how He is totally saved. It is about the belief in God that saves us. The belief in God is how it works. This is about the belief in God. This is how the belief in God becomes. It is about belief that totally works, that we are in belief, which we are totally about belief. This is what belief is, about that is about belief. It is about belief that is about belief that is about belief. This is about what belief is about, that is about belief. It is about belief that is about belief. This is totally about belief. This is about belief, which is about the world of belief, which is about the world of belief. The belief of the world is about the belief, of the world. This is about the belief of God. The belief of God is about the way you believe in God. This is about the belief in God. This is how God believes in God.

This is through others. This is about the others that are about the others. This is about the world that is about the world that is about the world. This is about the world that is about the world that is about the world. This is about the world that is about the world that is about the world. This is what the world is about the world that is about the world that is about the world. This is about the world that is about the world. This is about the world that is about the world that is about the world that is about the world. This is about the world that is about the world that is about the world. This is about the world that is about the world that is about the world. The world is about the world that is about the world that is about the world that is about the world that is about the world that is about the world. Belief is from above. Belief is from God only. He is the one that is the one that is the one. He is the chosen one. This is what God is about. He is about the chosen one known as the God of the universe. He is about salvation which is from the blood of the Lamb. This is about how the Lamb works and operates. He is a symbol from above. It is a real Lamb that saves us.

This is about the mercy that is from Jesus Christ. He has the Lamb do the real salvation for Him. He is about reality and how it is saved. This is about how Jesus Christ is about reality and how it is saved. This is about how the Lamb is about itself. This is about how the Lamb is about salvation. He is the author of the book of life. This is about how the Lamb is not a prophet or a saint. He is about the world. This is about the Lamb's mastery and focus. This is about gaining salvation for the chosen people. This is about the reality of salvation and how it works. This is about becoming what you want to become according to it. This is about how the world is about how the world is about how the world is about how the world is about how the world is about how the world is about how the world is about how the world is about how the world is about the world. This is about the world. This is about how the world is about the world that is about the world. This is about the world that is about the world. The world is about how the world is about how the world is about how the world is about deformation. This is about the world. This is about how the world is about how the world is about how the world that is about how the world is about how the world is about Riley Miller. He is a saved Christ writing to you. This is about salvation that is about from above. This is what salvation is about that is about salvation. This is about the Lamb that is about the Lamb that is about the Lamb that is about the Lamb that is about the Lamb. The Lamb's book of life has any Christian written down in it.

The Lamb's book of life was any name ever written down. This is anyone who accepted the free gift of life. This is about every name ever written down that is about Christ. This is what the Lamb's book of life is about. It is about names written down who understand the true meaning of Christ. This is those who worship the God of the Christian Bible. This God is Jesus Christ. He is about miracles and wonders. This is about how He gets along and how He gets along, is through the Book of Life.

This is every person. This is how every person is saved. This is through the Book of Life. This is how every person is saved. They all know the Christian meaning of Christ.

This is not what the beast wanted. He wanted to make everyone followers of him. This is about how he makes everyone a follower of Him. This is how Christ rules the world. This is about how Christ rules the world, and how He rules the world is through Christ. This means that there are false followers of the beast. This is about how He becomes that is about how He becomes. He becomes a follower of Himself.

This is not a trick. He is God. This is about forgiveness, and how He becomes with it. This is what the Antichrist cannot do. This is

Naturalization

what the world is about. It is about sin. This is
how it follows it.

The Antichrist is about sin. He is the
follower of reality. This is what reality is for him.
This is sin. The reality of sin is the Antichrist.
And, he tricks everyone into believing that. This
is why.

This is because he is a person. He is just a
person who walks the earth.

Wise Words
This is about the Antichrist and how he
survives. This is through the sacrifices to the
Lamb and the people who have died. This is
about knowing everything.

He knows everything. This is about the
Antichrist and how he survives. He survives
through the law. The Antichrist cannot be a
insane or weird person. He thrives off of love and
devotion. This is so he cannot have it.

This means he manipulates Christians who
cannot have love or devotion. This means he is
on fire for Jesus in a bad way. He is about ones
who you lost and how you hate them. He is about
falsehood and about practice. He is about cops
and their order. He is about you and me.
Deformation is real, and it is a process, to define
the real. This is how you define it to yourself.
Deformation is how you define it to all. The
process of deformation is about how you become,
and how you become, is about deformation. This
is about deformation and about how it works. It
works through love and devotion. This is because
it is a Christian theory. It is a theory in
Christianity that is about values. This is what it is
based around. It is based around domination. It is

a domination of values and words. This is how it has meaning. This is about a sane world of domination of values with words and phrases. This is about Satan's lies and how they are not accountable. This is with wise words, and phrases. This is about lies and deceit. This is only if you are a liar, and a deceiver. This is not about Satan's theory.

This is about Christ's theory. This is about Christ's theory that is about Christ. This is about how we work with Christ. This is on a personal basis. This is an everyday routine. This is how we stop the insanity of the Antichrist and his power. This is a needless agenda of demand. This is for Christ's love. This is all about Christ's love. This is about the needless agenda of Christ. Christ is how we love one another, which is obvious. If you are really interested in Christ, you would be interested in the Lamb.

This is a gentle topic. This is for deformation. This is about Christ, that is about Christ, that is about Christ. He is about ruling the world. He is about love, and how devotion works. He is not about abuse or denial. He is about the clean values of a clean slate. He is about the Lamb.

This is about His dark side. He doesn't have a dark side. He parties on the clean side. This is about His cleanliness and how He cherishes that. He is about world dominion and how He rules the world. He is about cherished and kind people.

He is about a servant. His name is Jesus Christ. He is a good man because He is God. He is about respect. He demands respect from all. This is how He demands respect. He demands

Naturalization

respect, because that is what He gets. This is about how all are different and the same too. This is about how we are all different. Yet, Christ loves all people. This is about how Christ is about Christ, which is about Christ, which is about Christ, which is about Christian values. This is about love, respect, and honor toward everyone. The Lamb is a perfect Christian who does no wrong. This is about how Christ rules. It is about the Antichrist. He is in fact insane.

Antichrists are Antichrists. They are non-Christians. They worship the mark of the beast. This is the number of sinners. This is how they work.

They plot against Christ day and night. They are in fraternities and sororities and the highest clubs in the world. The fact that Antichrists are Antichrists is the fact that God is a person. This means that people can see what is wrong. This only thing wrong is the Antichrist. He is insane.

Insane.
Insane.
Insane.
Insane.
Insane.
Insane.
Insane.
Insane.
Insane.
Insane.
Insane.
Insane.
Insane.
Insane.
Insane.

Naturalization

Insane.
Insane.
Insane.
Insane.
Insane.
Insane.
Insane.
Insane.
Insane.
Insane.
Insane.
Insane.
Insane.
Insane.
Insane.
Insane.
Insane.
Insane.
Insane.
Insane.
Insane.
Insane.
Insane.
Insane.
Insane.
Insane.
Insane.
Insane.
Insane.
Insane.
Insane.
Insane.
Insane.
Insane.

Naturalization

Insane.
Insane.
Insane.
Insane.
Insane.
Insane.
Insane.
Insane.
Insane.
Insane.
Insane.
Insane.
Insane.
Insane.
Insane.
Insane.
Insane.
Insane.
Insane.
Insane.
Insane.
Insane.
Insane.
Insane.
Insane.
Insane.
Insane.
Insane.
Insane.
Insane.
Insane.
Insane.
Insane.
Insane.
Insane.

Naturalization

Insane.
Insane.
Insane.
Insane.
Insane.
Insane.
Insane.
Insane.
Insane.
Insane.
Insane.
Insane.
Insane.
Insane.
Insane.
Insane.
Insane.
Insane.
Insane.
Insane.
Insane.

God is about seeing through the eyes of God. If we see through the eyes of God, we are saved. God sees through our eyes to see if we are saved. This is why we have difficulty viewing God. This is through a collide scope of magic and fascination. God has real eyes that see through us. We are all children of God, and what we see is through the children of God.

What we see is our reality. God is about seeing through the eyes of God.

Our reality is shaped with how we view God.

Naturalization

Our wonderful reality is shaped by demons and monsters. But, there is no test of strength for God. He is our wonderful counselor and prince of peace, Almighty God, Holy One.

He is about the view of the world, which is about His view. This is about the wonderful bewilderment of God, and how even Stephen Hawking can't understand it. This is about how non-Christians can't understand it only. These are what time has to offer. In Stephen Hawkings's wonderful book, "A Brief History of Time", He describes the God of the world and how He sees it.

It is through time that we cannot understand God. God is about time, and time is about God.

I like nice people.

I wrote a thesis paper on America and how it is everyone's favorite.

God sees through the eyes of God, and we are all saved. This is just trials and tribulations we go through. A Christ would fully explain what He has to go through to become a Christian.

What we have to go through is just trials and tribulations. Everyone is saved in the end.

I even put my honor on it. This is about Christ and salvation. He has promised to save us all.

This is a mission of God. If Stephen Hawking can write a "Brief History of Time", about God and how He sees time then anyone can be saved.

God's Genius

Naturalization

God is smart. This is what He is smart for and about. He is about smart things for wonderful proportions. He is about God and what God is about is how he is smart.

God is a wonderful counselor of magic and fascination.

The modern day has worshipped God as a powerful ruler. This is why he created the world. This is because it is His. It was already His. He is a ruler who is Mighty and Strong. He is about bigger and better things. He is about the Almighty God, of the world.

Yet, Jesus Christ is His Son. Jesus was the forgiver of all. He washed clean His salvation when He died on the cross. He washes away our sins. This is eternally.

Jesus died on the cross to make all men His. His personality saved all of us. This is why He is a Lamb. The Lamb harms no one. This is about the Lamb's story and about poverty. He was a God. He was Jesus Christ. And, He has yet to come. All we know of Him about is through stories. He is about salvation.

This is about how we become.

This is through Christ only.

This is even how the Antichrist becomes.

So, in the end, we are all saved. We are really all humans, and there is an allegory to everything known as the Bible. This Bible, is known as the Book, that is the Immaculate Conception. This is insane, sometimes, because of Jesus Christ. This is how smart the Antichrist is and how his number is the number of man. This is about God's love. This is toward all of us.

He comes in the end to save all of us. This is why:

Naturalization

1. God is just
2. God is good.
3. God is powerful
4. God is Forgiveness
5. God is love
6. God is Grace
7. God is Awesome.

God is what God is. This is really what God is. He is a Wonderful Counselor. He would not have created the world if we were not all saved. This is His initial design. It is not just the Bible's. He is the counselor of all. This is what God is about. He is about saving all. He is about creation and how it works. He is a wonderful God.

Meet Jesus Christ

This is about how Jesus Christ rules the world. Jesus Christ rules with mercy and compassion. He is the chosen one. He is the ruler of all that is human.
This is why we are all saved. If God is omniscient, and omnipotent, and omnipresent, then He obviously had a mission to save the world. This is what a fact is.

If He did not, then there would be no Heaven. If there is justice in the world, then we are all saved. This is why my book is superior to what people think about it. This is because of Jesus' Divine Plan. This is about Him. He is a savior.

This is about how the Lamb copes with sin. He is capable of being delivered from evil. This is all of Him. This is all His followers.

Naturalization

This is why He is God. He is capable of being God. This is why.
God is:
1. Omniscient
2. Omnipresent
3. Omnipotent
4. Ever Graceful
5. Ever Loving
6. Ever Good
7. Ever Graceful

God is good. God is about the true following. God is about the Great Following. He is capable of moving mountains. This is about Jesus Christ. Jesus is about forgiveness, and about being capable of being good. Jesus is about all of creation and all of creation is about forgiveness. This is about being capable of being forgiven. This is about the wonderful creation. This is about God's Divine Immaculate Plan. This is about being able to be forgiven for.

This is not about being condemned. This is about the beautiful occupation. This is about life and how we all are about life. This is about how we are about God. God is about God's Plan and about fulfilling it. This is even if we are a Christian. This is how the world is ruled. This is about God's Immaculate Conception Plan. This is about God's Immaculate Conception Plan. He is about God. This is how God understands others. He is forgives by understanding others when they understand Him. God is about progress and forgiving others. This is if you are me.

If you are God you would understand it differently.

Naturalization

He just sees us.
He becomes us.

This is about the world that is about the
world that is about the world that is about the
world that is about the world that is about the
world that is about the world that is about God.
This is about the world and how the world works
and how the world works and how the world
works and how the world works and how the
world works and how it works. This is about the
world that is about the world that is about the
world that is about the world that is about the
world that is about the world that is about the
world that is about the world that is about the
world that is about the world that is about the
world that is about the world that is about the
world that is about the world that is about the
world that is about the world that is about the
world that is about the world that is about the
world that is about the world that is about the
world that is about the world that is about the
world that is about the world that is about the
world that is about the world that is about the
world that is about the world that is about the
world that is about the world that is about the
world that is about the world that is about the
world that is about the world that is about the
world that is about the world that is about the
world that is about the world that is about the
world that is about the world that is about the
world that is about the world that is about the
world that is about the world that is about the
world that is about the world that is about the
world that is about the world that is about the
world that is about the world that is about the
world that is about the world that is about the

Naturalization

world that is about the world that is about the
world that is about the world that is about the
world that is about the world that is about the
world that is about the world that is about the
world that is about the world that is about the
world that is about the world that is about the
world that is about the world that is about the
world that is about the world that is about the
world that is about the world that is about the
world that is about the world that is about the
world that is about the world that is about the
world that is about the world that is about the
world that is about the world that is about the
world that is about the world that is about the
world that is about the world that is about the
world that is about the world that is about the
world that is about the world that is about the
world that is about the world that is about the
world that is about the world that is about the
world that is about the world that is about the
world that is about the world that is about the
world that is about the world that is about the
world that is about the world that is about the
world that is about the world that is about the
world that is about the world that is about the
world that is about the world that is about the
world that is about the world that is about the
world that is about the world that is about the
world that is about the world that is about the
world that is about the world that is about the
world that is about the world that is about the
world that is about the world that is about the
world that is about the world that is about the
world that is about the world that is about the
world that is about the world that is about the
world that is about the world that is about the
world that is about the world that is about the

Naturalization

world that is about the world that is about the
world that is about the world that is about the
world that is about the world that is about the
world that is about the world that is about the
world that is about the world that is about the
world that is about the world that is about the
world that is about the world that is about the
world that is about the world that is about the
world that is about the world that is about the
world that is about the world that is about the
world that is about the world that is about the
world that is about the world that is about the
world that is about the world that is about the
world that is about the world that is about the
world that is about the world that is about the
world that is about the world that is about the
world that is about the world that is about the
world that is about the world that is about the
world that is about the world that is about the
world that is about the world that is about the
world that is about the world that is about the
world that is about the world that is about the
world that is about the world that is about the
world that is about the world that is about the
world that is about the world that is about the
world that is about the world that is about the
world that is about the world that is about the
world that is about the world that is about the
world that is about the world that is about the
world that is about the world that is about the
world that is about the world that is about the
world that is about the world that is about the
world that is about the world that is about the
world that is about the world that is about the
world that is about the world that is about the
world that is about the world that is about the

Naturalization

world that is about the world that is about the
world that is about the world that is about the
world that is about the world that is about the
world that is about the world that is about the
world that is about the world that is about the
world that is about the world that is about the
world that is about the world that is about the
world that is about the world that is about the
world that is about the world that is about the
world that is about the world that is about the
world that is about the world that is about the
world that is about the world that is about the
world that is about the world that is about the
world that is about the world that is about the
world that is about the world that is about the
world that is about the world that is about the
world that is about the world that is about the
world that is about the world that is about the
world that is about the world that is about the
world that is about the world that is about the
world that is about the world that is about the
world that is about the world that is about the
world that is about the world that is about the
world that is about the world that is about the
world that is about the world that is about the
world that is about the world that is about the
world that is about the world that is about the
world that is about the world that is about the
world that is about the world that is about the
world that is about the world that is about the
world that is about the world that is about the
world that is about the world that is about the
world that is about the world that is about the
world that is about the world that is about the
world that is about the world that is about the
world that is about the world that is about the

Naturalization

world that is about the world that is about the
world that is about the world that is about the
world that is about the world that is about the
world that is about the world that is about the
world that is about the world that is about the
world that is about the world that is about the
world that is about the world that is about the
world that is about the world that is about the
world that is about the world that is about the
world that is about the world that is about the
world that is about the world that is about the
world that is about the world that is about the
world that is about the world that is about the
world that is about the world that is about the
world that is about the world that is about the
world that is about the world that is about the
world that is about the world that is about the
world that is about the world that is about the
world that is about the world that is about the
world that is about the world that is about the
world that is about the world that is about the
world that is about the world that is about the
world that is about the world that is about the
world that is about the world that is about the
world that is about the world that is about the
world that is about the world that is about the
world that is about the world that is about the
world that is about the world that is about the
world that is about the world that is about the
world that is about the world that is about the
world that is about the world that is about the
world that is about the world that is about the
world that is about the world that is about the
world that is about the world that is about the
world that is about the world that is about the
world that is about the world that is about the
world that is about the world that is about the

Naturalization

world that is about the world that is about the
world that is about the world that is about the
world that is about the world that is about the
world that is about the world that is about the
world that is about the world that is about the
world that is about the world that is about the
world that is about the world that is about the
world that is about the world that is about the
world that is about the world that is about the
world that is about the world that is about the
world that is about the world that is about the
world that is about the world that is about the
world that is about the world that is about the
world that is about the world that is about the
world that is about the world that is about the
world that is about the world that is about the
world that is about the world that is about the
world that is about the world that is about the
world that is about the world that is about the
world that is about the world that is about the
world that is about the world that is about the
world that is about the world that is about the
world that is about the world that is about the
world that is about the world that is about the
world that is about the world that is about the
world that is about the world that is about the
world that is about the world that is about the
world that is about the world that is about the
world that is about the world that is about the
world that is about the world that is about the
world that is about the world that is about the
world that is about the world that is about the
world that is about the world that is about the
world that is about the world that is about the
world that is about the world that is about the

Naturalization

world that is about the world that is about the
world that is about the world that is about the
world that is about the world that is about the
world that is about the world that is about the
world that is about the world that is about the
world that is about the world that is about the
world that is about the world that is about the
world that is about the world that is about the
world that is about the world that is about the
world that is about the world that is about the
world that is about the world that is about the
world that is about the world that is about the
world that is about the world that is about the
world that is about the world that is about the
world that is about the world that is about the
world that is about the world that is about the
world that is about the world that is about the
world that is about the world that is about the
world that is about the world that is about the
world that is about the world that is about the
world that is about the world that is about the
world that is about the world that is about the
world that is about the world that is about the
world that is about the world that is about the
world that is about the world that is about the
world that is about the world that is about the
world that is about the world that is about the
world that is about the world that is about the
world that is about the world that is about the
world that is about the world that is about the
world that is about the world that is about the
world that is about the world that is about the
world that is about the world that is about the
world that is about the world that is about the
world that is about the world that is about the
world that is about the world that is about the

Naturalization

world that is about the world that is about the
world that is about the world that is about the
world that is about the world that is about the
world that is about the world that is about the
world that is about the world that is about the
world that is about the world that is about the
world that is about the world that is about the
world that is about the world that is about the
world that is about the world that is about the
world that is about the world that is about the
world that is about the world that is about the
world that is about the world that is about the
world that is about the world that is about the
world that is about the world that is about the
world that is about the world that is about the
world that is about the world that is about the
world that is about the world that is about the
world that is about the world that is about the
world that is about the world that is about the
world that is about the world that is about the
world that is about the world that is about the
world that is about the world that is about the
world that is about the world that is about the
world that is about the world that is about the
world that is about the world that is about the
world that is about the world that is about the
world that is about the world that is about the
world that is about the world that is about the
world that is about the world that is about the
world that is about the world that is about the
world that is about the world that is about the
world that is about the world that is about the
world that is about the world that is about the
world that is about the world that is about the
world that is about the world that is about the
world that is about the world that is about the
world that is about the world that is about the
world that is about the world that is about the

Naturalization

world that is about the world that is about the
world that is about the world that is about the
world that is about the world that is about the
world that is about the world that is about the
world that is about the world

that is about the world that is about the world that
is about the world that is about the world that is
about the world that is about the world that is
about the world that is about the world that is
about the world that is about the world that is
about the world that is about the world that is
about the world that is about the world that is
about the world that is about the world that is
about the world that is about the world that is
about the world that is about the world that is
about the world that is about the world that is
about the world that is about the world that is
about the world that is about the world that is
about the world that is about the world that is
about the world that is about the world that is
about the world that is about the world that is
about the world that is about the world that is
about the world that is about the world that is
about the world that is about the world that is
about the world that is about the world that is
about the world that is about the world that is
about the world that is about the world that is
about the world that is about the world that is
about the world that is about the world that is
about the world that is about the world that is
about the world that is about the world that is
about the world that is about the world that is
about the world that is about the world that is
about the world that is about the world that is
about the world that is about the world that is
about the world that is about the world that is

Naturalization

about the world that is about the world that is
about the world that is about the world that is
about the world that is about the world that is
about the world that is about the world that is
about the world that is about the world that is
about the world that is about the world that is
about the world that is about the world that is
about the world that is about the world that is
about the world that is about the world that is
about the world that is about the world that is
about the world that is about the world that is
about the world that is about the world that is
about the world that is about the world that is
about the world that is about the world that is
about the world that is about the world that is
about the world that is about the world that is
about the world that is about the world that is
about the world that is about the world that is
about the world that is about the world that is
about the world that is about the world that is
about the world that is about the world that is
about the world that is about the world that is
about the world that is about the world that is
about the world that is about the world that is
about the world that is about the world that is
about the world that is about the world that is
about the world that is about the world that is
about the world that is about the world that is
about the world that is about the world that is
about the world that is about the world that is
about the world that is about the world that is
about the world that is about the world that is
about the world that is about the world that is
about the world that is about the world that is
about the world that is about the world that is
about the world that is about the world that is
about the world that is about the world that is

Naturalization

about the world that is about the world that is
about the world that is about the world that is
about the world that is about the world that is
about the world that is about the world that is
about the world that is about the world that is
about the world that is about the world that is
about the world that is about the world that is
about the world that is about the world that is
about the world that is about the world that is
about the world that is about the world that is
about the world that is about the world that is
about the world that is about the world that is
about the world that is about the world that is
about the world that is about the world that is
about the world that is about the world that is
about the world that is about the world that is
about the world that is about the world that is
about the world that is about the world that is
about the world that is about the world that is
about the world that is about the world that is
about the world that is about the world that is
about the world that is about the world that is
about the world that is about the world that is
about the world that is about the world that is
about the world that is about the world that is
about the world that is about the world that is
about the world that is about the world that is
about the world that is about the world that is
about the world that is about the world that is
about the world that is about the world that is
about the world that is about the world that is
about the world that is about the world that is
about the world that is about the world that is
about the world that is about the world that is
about the world that is about the world that is
about the world that is about the world that is
about the world that is about the world that is

Naturalization

about the world that is about the world that is
about the world that is about the world that is
about the world that is about the world that is
about the world that is about the world that is
about the world that is about the world that is
about the world that is about the world that is
about the world that is about the world that is
about the world that is about the world that is
about the world that is about the world that is
about the world that is about the world that is
about the world that is about the world that is
about the world that is about the world that is
about the world that is about the world that is
about the world that is about the world that is
about the world that is about the world that is
about the world that is about the world that is
about the world that is about the world that is
about the world that is about the world that is
about the world that is about the world that is
about the world that is about the world that is
about the world that is about the world that is
about the world that is about the world that is
about the world that is about the world that is
about the world that is about the world that is
about the world that is about the world that is
about the world that is about the world that is
about the world that is about the world that is
about the world that is about the world that is
about the world.

The End is the end as the end. This is what the
end means. This is just like the end of all. This is
about the end of the world that is about the end of
the world. This is what happens when you do not
become with Jesus. This is about what God is
about that is about the one true God. His name is
Jesus Christ. This is about Jesus. This is about
what Jesus did that is about Jesus Christ. He
saved all of humanity. This is about how He

Naturalization

saved all of humanity. This is by reaching out to people and caring for them. This is about how the world that is about the world is about the world. This is about what God is about that is about what God is about. This is about forgiving when you are forgiven. This is about the world that is about the world that is about the world that is about the universe that is about the world that is about the universe. This is my conclusion. Thought is different than what thought is about. This is about what thought is about. This is about great and wonderful things that are about what thought is about. This is about what thought is about that is about what the world is about that is about the world. This is about what God is about and what God is about is about what God is about. God is about God that is about God. God is about God. This is about God and how God is about how God is about how God is about how God is about how God is about how God is about how God is about how God is about how God is about how God is about how God is about God. This is about how God is about God. This is about Jesus Christ.

Naturalization

Naturalization

<u>FROM THE AUTHOR'S DESK</u>

I come from a very close Christian family. I am 30. I am white. I am a real big Christian. I am male. I have written over 86 books, as of the age of 30. I am trying to start my own construction of government, with the "New World Order."

You can visit me on the Facebook.com page under "Riley Miller". My email address is RPMDallas1@gmail.com. I reside and live in Dallas, Texas. This is in the Preston Hollow region.

5800 Royal Lane Private Drive
10711 Villager Road, Unit C
75230.

Naturalization

www.ingramcontent.com/pod-product-compliance
Lightning Source LLC
Chambersburg PA
CBHW071520180526
45171CB00002B/326